Advances in Classical Field Theory

Asher Yahalom

Ariel University Center of Samaria, Ariel 40700, Israel

TO MY WIFE HADASS

AND MY CHILDREN URI, NURIT AND ASAF

Contents

Preface

Classical field theory is used by physicists to describe a wide variety of physical phenomena. Those include electromagnetism, fluid dynamics, gravitation and quantum mechanics. The central entity of field theory is the field which is usually a multi component function of space and time. Those multi component functions are usually grouped together as vector fields as in the case of electromagnetic theory and fluid dynamics, in other cases they are grouped as tensors as in theory of gravitation and yet in other cases they are grouped as complex functions as in the case of quantum mechanics. In order to know the value of the field one needs to solve a set of coupled partial differential equations with given boundary and initial conditions. Both the equations and the boundary and initial conditions are in most cases available from a variational principle through an adequate Lagrangian, although in some cases this is less straight forward as in the case of fluid dynamics. An appropriate field theory satisfies two main requirements: First it should describe correctly the relevant physical phenomena and second the theory should describe the phenomena with a minimal number of functions, equations and assumptions. Advances in classical theory have thus two principal meanings. One meaning is improvement of a known theory in the sense that the new theory proposed now describes correctly a wider class of physical phenomena. A second meaning is that a given physical phenomena is described by less functions, equations or assumptions than was needed before.

This book is aimed at introducing recent advances in classical field theory of both meanings. The book will cover recent advances in electromagnetism, fluid dynamics, gravitation and quantum mechanics. Not all advances will be covered but rather a selected group.

A final note regarding the meaning of the word "classical" in the present context. In classical field theory the fields are multi component functions, in contrast to "quantum" field theories in which the fields are multi component operators. The bridge between classical and quantum fields is the process of "second quantization" or through the Feynman path integral. The quantum field theory is usually used to describe elementary particle phenomena and will not be discussed here.

I would like to thank all contributors to the present book, those include (in alphabetical order): Prof. Robert Englman, Dr. Yuri Lurie, Dr. Gad A. Pinhasi, Prof. Yosef Pinhasi and Dr. Marcelo Schiffer. The name of each of the contributors is indicated on the chapter which he contributed. Chapters in which no name appear were authored solely by the editor of this book.

Last but not least thanks are due to Mr. Netanel Gabdank for his meticulous preparation of the index of this book and to the editorial director Dr. Matthew Honan and Ms. Bushra Siddiqui of Bentham Science Publishers for prolific collaboration.

Asher Yahalom
Ariel University Center of Samaria
Israel

Contributors

Asher Yahalom
Ariel University Center of Samaria, Ariel 40700, Israel

Robert Englman
Department of Physics and Applied Mathematics, Soreq NRC, Yavne 81800, Israel

Yosef Pinhasi
Ariel University Center of Samaria, P.O.B. 3, Ariel 40700 Israel

Yuri Lurie
Ariel University Center of Samaria, P.O.B. 3, Ariel 40700 Israel

Gad A. Pinhasi
Ariel University Center of Samaria, P.O.B. 3, Ariel 40700 Israel

Marcelo Schiffer
Ariel University Center of Samaria, P.O.B. 3, Ariel 40700 Israel

CHAPTER 1

Advances in the Field Theory of Flows

Abstract

We introduce a three independent-functions field formalism for stationary and non-stationary barotropic flows. This is less than the four variables which appear in the standard equations of fluid dynamics which are the velocity field $\vec{v}$ and the density ρ.

1.1 Introduction to the Field Theory of Barotropic Fluid Dynamics

The formulation of barotropic fluid dynamics as a field theory, that is deriving its equations from a Lagrangian density through a variational principle, is now well known. A four function variational formulation of Eulerian barotropic fluid dynamics was derived by Davydov [1], but since the work was written in Russian, it was unknown in the west. Lagrangian fluid dynamics (as opposed to Eulerian fluid dynamics) was formulated through a variational principle by Eckart [2]. Initial western attempts to formulate Eulerian fluid dynamics in terms of a variational principle, were described by Herivel [3], Serrin [4], Lin [5]. However, the variational principles developed by the above authors were very cumbersome containing quite a few "Lagrange multipliers" and "potentials ". The total number of independent functions in the

Asher Yahalom (Ed)

above formulations ranges from seven to eleven which considerably exceeds the four functions appearing in the Eulerian and continuity equations of a barotropic flow. Therefore those approaches do not have any practical use or application. Seliger & Whitham [6] have developed a variational formalism which can be shown to depend on only four variables for barotropic flow and is equivalent to the formulation of Davydov [1]. Lynden-Bell & Katz [7] have described a variational principle in terms of two functions the load λ (to be described below) and density ρ. However, their formalism contains an implicit definition for the velocity $\vec{v}$ such that one is required to solve a partial differential equation in order to obtain both $\vec{v}$ in terms of ρ and λ as well as its variations . Much the same criticism holds for their general variational for non-barotropic flows [8]. Yahalom & Lynden-Bell [9] have overcome this limitation by paying the price of adding an additional single function. Their formalism allows arbitrary variations and the definition of $\vec{v}$ is explicit.

This chapter will describe the advancements in the description of barotropic fluid dynamics as a field theory. Emphasis will be given on the reduction of the degrees of freedom which is achieved through this formulation. Anticipated applications for this study include both stability analysis of known fluid dynamics configurations and designing efficient numerical schemes for integrating the equations of fluid dynamics [10, 11, 12, 13, 14, 15, 16, 17, 18, 19].

The plan of the following sections is as follows: We will review the basic equations of Eulerian fluid dynamics and give a somewhat different derivation of the four function variational principle of Davydov and Seliger & Whitham [1, 6]. Then we will describe the three function variational principle for non-stationary fluid dynamics [9]. A discussion of the fluid topological constants will follow. Finally we will give a different variational principle for stationary fluid dynamics [9].

1.2 Variational principle of non-stationary fluid dynamics

Barotropic Eulerian fluids can be described in terms of four functions the velocity $\vec{v}$ and density ρ. Those functions need to satisfy the continuity and Euler equations:

$$\frac{\partial \rho}{\partial t} + \vec{\nabla} \cdot (\rho \vec{v}) = 0 \tag{1.1}$$

$$\frac{d\vec{v}}{dt} = \frac{\partial \vec{v}}{\partial t} + (\vec{v} \cdot \vec{\nabla})\vec{v} = -\frac{\vec{\nabla} p(\rho)}{\rho} \tag{1.2}$$

In which the pressure $p(\rho)$ is assumed to be a given function of the density . Taking the curl of equation (1.2) will lead to:

$$\frac{\partial \vec{\omega}}{\partial t} = \vec{\nabla} \times (\vec{v} \times \vec{\omega}) \tag{1.3}$$

in which:

$$\vec{\omega} = \vec{\nabla} \times \vec{v} \tag{1.4}$$

is the vorticity . Equation (1.3) describes the fact that the vorticity lines are "frozen" within the Eulerian flow [1].

A very simple variational principle for non-stationary fluid dynamics was described by Davydov and Seliger & Whitham [1, 6] and is brought here mainly for completeness using a slightly different derivation than the one appearing in the original paper. This will serve as a starting point for the next section in which we will show how the variational principle can be simplified further. Consider the action:

$$\begin{aligned} A &\equiv \int \mathcal{L} d^3x dt \\ \mathcal{L} &\equiv \mathcal{L}_1 + \mathcal{L}_2 \\ \mathcal{L}_1 &\equiv \rho(\frac{1}{2}\vec{v}^2 - \varepsilon(\rho)), \qquad \mathcal{L}_2 \equiv \nu[\frac{\partial \rho}{\partial t} + \vec{\nabla} \cdot (\rho \vec{v})] - \rho\alpha\frac{d\beta}{dt} \end{aligned} \tag{1.5}$$

in which $\varepsilon(\rho)$ is the specific internal energy . Obviously ν, α are Lagrange multipliers which were inserted in such a way that the variational principle will yield the following equations:

$$\begin{aligned} \frac{\partial \rho}{\partial t} + \vec{\nabla} \cdot (\rho \vec{v}) = 0 \\ \rho \frac{d\beta}{dt} = 0 \end{aligned} \tag{1.6}$$

Provided ρ is not null those are just the continuity equation (1.1) and the conditions that β is comoving. Let us take an arbitrary variational derivative

[1]The most general vortical flux and mass preserving flows that may be attributed to vortex lines were found in [20]

of the above action with respect to $\vec{v}$, this will result in:

$$\delta_{\vec{v}} A = \int d^3x dt \rho \delta \vec{v} \cdot [\vec{v} - \vec{\nabla}\nu - \alpha \vec{\nabla}\beta] + \oint d\vec{S} \cdot \delta \vec{v} \rho \nu \tag{1.7}$$

Provided that the above boundary term vanishes, as in the case of astrophysical flows for which $\rho = 0$ on the free flow boundary, or the case in which the fluid is contained in a vessel which induces a no flux boundary condition $\delta \vec{v} \cdot \hat{n} = 0$ ($\hat{n}$ is a unit vector normal to the boundary), $\vec{v}$ must have the following form:

$$\vec{v} = \hat{\vec{v}} \equiv \alpha \vec{\nabla}\beta + \vec{\nabla}\nu \tag{1.8}$$

this is nothing but Clebsch representation of the flow field (see for example [2], [21, page 248]). Let us now take the variational derivative with respect to the density ρ, we obtain:

$$\delta_\rho A = \int d^3x dt \delta\rho [\frac{1}{2}\vec{v}^2 - w - \frac{\partial \nu}{\partial t} - \vec{v} \cdot \vec{\nabla}\nu] + \oint d\vec{S} \cdot \vec{v} \delta \rho \nu + \int d^3x \nu \delta\rho |_{t_0}^{t_1} \tag{1.9}$$

in which $w = \frac{\partial(\rho\varepsilon)}{\partial\rho}$ is the specific enthalpy . Hence provided that $\delta\rho$ vanishes on the boundary of the domain and in initial and final times the following equation must be satisfied:

$$\frac{d\nu}{dt} = \frac{1}{2}\vec{v}^2 - w \tag{1.10}$$

Finally we have to calculate the variation with respect to β this will lead us to the following results:

$$\delta_\beta A = \int d^3x dt \delta\beta [\frac{\partial(\rho\alpha)}{\partial t} + \vec{\nabla} \cdot (\rho\alpha\vec{v})] - \oint d\vec{S} \cdot \vec{v}\rho\alpha\delta\beta - \oint d\vec{\Sigma} \cdot \vec{v}\rho\alpha\delta\beta - \int d^3x \rho\alpha\delta\beta|_{t_0}^{t_1} \tag{1.11}$$

Hence choosing $\delta\beta$ in such a way that the temporal and spatial boundary terms vanish including boundary conditions on a "cut" Σ in the case that the function β is not single valued, in the above integral will lead to the equation:

$$\frac{\partial(\rho\alpha)}{\partial t} + \vec{\nabla}\cdot(\rho\alpha\vec{v}) = 0 \tag{1.12}$$

Notice that the variation on the cut will only vanish if the variation of β is single valued or a "Kutta" condition is satisfied [15] on the cut, that is the velocity field is parallel to the cut. Using the continuity equation (1.1) this will lead to the equation:

$$\rho\frac{d\alpha}{dt} = 0 \tag{1.13}$$

Hence for $\rho \neq 0$ both α and β are comoving coordinates. Since the vorticity can be easily calculated from equation (1.8) to be:

$$\vec{\omega} = \vec{\nabla}\times\vec{v} = \vec{\nabla}\alpha\times\vec{\nabla}\beta \tag{1.14}$$

Calculating $\frac{\partial\vec{\omega}}{\partial t}$ in which ω is given by equation (1.14) and taking into account both equation (1.13) and equation (1.6) will yield equation (1.3).

1.2.1 Euler's equations

We shall now show that a velocity field given by equation (1.8), such that the functions α, β, ν satisfy the corresponding equations (1.6,1.10,1.13) must satisfy Euler's equations. Let us calculate the material derivative of $\vec{v}$:

$$\frac{d\vec{v}}{dt} = \frac{d\vec{\nabla}\nu}{dt} + \frac{d\alpha}{dt}\vec{\nabla}\beta + \alpha\frac{d\vec{\nabla}\beta}{dt} \tag{1.15}$$

It can be easily shown that:

$$\begin{aligned}
\frac{d\vec{\nabla}\nu}{dt} &= \vec{\nabla}\frac{d\nu}{dt} - \vec{\nabla}v_k\frac{\partial\nu}{\partial x_k} = \vec{\nabla}(\frac{1}{2}\vec{v}^2 - w) - \vec{\nabla}v_k\frac{\partial\nu}{\partial x_k} \\
\frac{d\vec{\nabla}\beta}{dt} &= \vec{\nabla}\frac{d\beta}{dt} - \vec{\nabla}v_k\frac{\partial\beta}{\partial x_k} = -\vec{\nabla}v_k\frac{\partial\beta}{\partial x_k}
\end{aligned} \tag{1.16}$$

In which x_k is a Cartesian coordinate and a summation convention is assumed. Inserting the result from equations (1.16) into equation (1.15) yields:

$$\begin{aligned}
\frac{d\vec{v}}{dt} &= -\vec{\nabla}v_k(\frac{\partial\nu}{\partial x_k} + \alpha\frac{\partial\beta}{\partial x_k}) + \vec{\nabla}(\frac{1}{2}\vec{v}^2 - w) \\
&= -\vec{\nabla}v_k v_k + \vec{\nabla}(\frac{1}{2}\vec{v}^2 - w) = -\frac{\vec{\nabla}p}{\rho}
\end{aligned} \tag{1.17}$$

This proves that the Euler equations can be derived from the action given in equation (1.5) and hence all the equations of fluid dynamics can be derived from the above action without restricting the variations in any way. Taking the curl of equation (1.17) will lead to equation (1.3).

1.2.2 Simplified action

The reader of this chapter might argue that the author has introduced unnecessary complications to the theory of fluid dynamics by adding three more functions α, β, ν to the standard set $\vec{v}, \rho$. In the following we will show that this is not so and the action given in equation (1.5) in a form suitable for a pedagogic presentation can indeed be simplified. It is easy to show that the Lagrangian density appearing in equation (1.5) can be written in the form:

$$\begin{aligned} \mathcal{L} &= -\rho[\frac{\partial \nu}{\partial t} + \alpha \frac{\partial \beta}{\partial t} + \varepsilon(\rho)] + \frac{1}{2}\rho[(\vec{v} - \hat{\vec{v}})^2 - \hat{\vec{v}}^2] \\ &+ \frac{\partial(\nu\rho)}{\partial t} + \vec{\nabla}\cdot(\nu\rho\vec{v}) \end{aligned} \tag{1.18}$$

In which $\hat{\vec{v}}$ is a shorthand notation for $\vec{\nabla}\nu + \alpha\vec{\nabla}\beta$ (see equation (1.8)). Thus $\mathcal{L}$ has three contributions:

$$\begin{aligned} \mathcal{L} &= \hat{\mathcal{L}} + \mathcal{L}_{\vec{v}} + \mathcal{L}_{boundary} \\ \hat{\mathcal{L}} &\equiv -\rho[\frac{\partial \nu}{\partial t} + \alpha \frac{\partial \beta}{\partial t} + \varepsilon(\rho) + \frac{1}{2}(\vec{\nabla}\nu + \alpha\vec{\nabla}\beta)^2] \\ \mathcal{L}_{\vec{v}} &\equiv \frac{1}{2}\rho(\vec{v} - \hat{\vec{v}})^2 \\ \mathcal{L}_{boundary} &\equiv \frac{\partial(\nu\rho)}{\partial t} + \vec{\nabla}\cdot(\nu\rho\vec{v}) \end{aligned} \tag{1.19}$$

The only term containing $\vec{v}$ is $\mathcal{L}_{\vec{v}}$, it can easily be seen that this term will lead, after we nullify the variational derivative, to equation (1.8) but will otherwise have no contribution to other variational derivatives. Notice that the term $\mathcal{L}_{boundary}$ contains only complete partial derivatives and thus can not contribute to the equations although it can change the boundary conditions. Hence we see that equations (1.6), equation (1.10) and equation (1.13) can be derived using the Lagrangian density $\hat{\mathcal{L}}$ in which $\hat{\vec{v}}$ replaces $\vec{v}$ in the relevant equations. Notice, however, that some of the functions appearing in our formalism (α, β, ν) may turn out to be non single valued. In this case suitable "cuts" and boundary conditions on the cuts need to be introduced. In

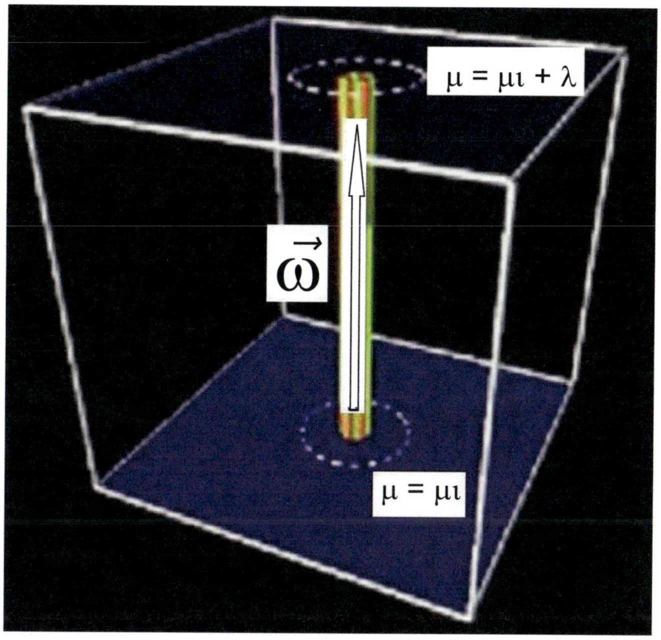

Figure 1.1: A thin tube surrounding a vortex line

particular it will be shown that ν is not single valued for flows with non-zero helicity . Furthermore, after integrating the four equations (1.6,1.10,1.13) we can insert the potentials α, β, ν into equation (1.8) to obtain the physical velocity $\vec{v}$. Hence, the general barotropic fluid dynamics problem is changed such that instead of solving the four equations (1.1,1.2) we need to solve an alternative set which can be derived from the Lagrangian density $\hat{\mathcal{L}}$.

1.2.3 The inverse problem

In the previous subsection we have shown that given a set of functions α, β, ν satisfying the set of equations described in the previous subsections, one can insert those functions into equation (1.8) and equation (1.14) to obtain the physical velocity $\vec{v}$ and vorticity $\vec{\omega}$. In this subsection we will address the inverse problem that is, suppose we are given the quantities $\vec{v}$ and ρ how can one calculate the potentials α, β, ν? The treatment in this section will follow closely (with minor changes) the discussion given by Lynden-Bell & Katz [7] and is given here for completeness.

Consider a thin tube surrounding a vortex line as described in figure 1.1, the vorticity flux contained within the tube which is equal to the circulation around the tube is:

$$\Delta\Phi = \int \vec{\omega} \cdot d\vec{S} = \oint \vec{v} \cdot d\vec{r} \quad (1.20)$$

and the mass contained with the tube is:

$$\Delta M = \int \rho d\vec{l} \cdot d\vec{S} \quad (1.21)$$

in which dl is a length element along the tube . Since the vortex lines move with the flow by virtue of equation (1.3) both the quantities $\Delta\Phi$ and ΔM are conserved and since the tube is thin we may define the conserved load :

$$\lambda = \frac{\Delta M}{\Delta\Phi} = \oint \frac{\rho}{\omega} dl \quad (1.22)$$

in which the above integral is performed along the field line. Obviously the parts of the line which go out of the flow to regions in which $\rho = 0$ have a null contribution to the integral. Since λ is conserved it satisfies the equation:

$$\frac{d\lambda}{dt} = 0. \quad (1.23)$$

By construction surfaces of constant load move with the flow and contain vortex lines. Hence the gradient to such surfaces must be orthogonal to the field line:

$$\vec{\nabla}\lambda \cdot \vec{\omega} = 0 \quad (1.24)$$

Now consider an arbitrary comoving point on the vortex line and donate it by i, and consider an additional comoving point on the vortex line and donate it by r. The integral:

$$\mu(r) = \int_i^r \frac{\rho}{\omega} dl + \mu(i) \quad (1.25)$$

is also a conserved quantity which we may denote following Lynden-Bell & Katz [7] as the generalized metage . $\mu(i)$ is an arbitrary number which can be chosen differently for each vortex line. By construction:

$$\frac{d\mu}{dt} = 0. \quad (1.26)$$

Also it is easy to see that by differentiating along the vortex line we obtain:

$$\vec{\nabla}\mu \cdot \vec{\omega} = \rho \quad (1.27)$$

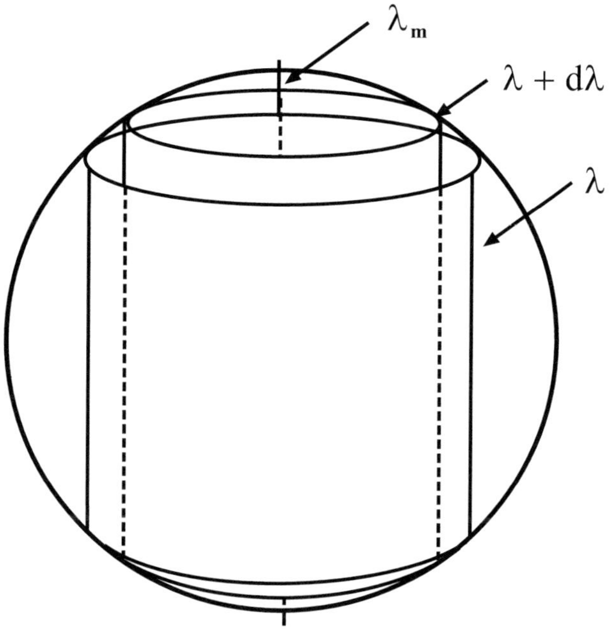

Figure 1.2: Surfaces of constant load

At this point we have two comoving coordinates of flow , namely λ, μ obviously in a three dimensional flow we also have a third coordinate. However, before defining the third coordinate we will find it useful to work not directly with λ but with a function of λ. Now consider the vortical flux $\Phi(\lambda)$ within a surface of constant load as described in figure 1.2 (the figure was given by Lynden-Bell & Katz [7]). The flux is a conserved quantity and depends only on the load λ of the surrounding surface. Now we define the quantity:

$$\alpha = \frac{\Phi(\lambda)}{2\pi} = \frac{C(\lambda)}{2\pi} \tag{1.28}$$

$C(\lambda)$ is the circulation along lines on this surface. Obviously α satisfies the equations:

$$\frac{d\alpha}{dt} = 0, \qquad \vec{\omega} \cdot \vec{\nabla}\alpha = 0 \tag{1.29}$$

Let us now define an additional comoving coordinate β^* since $\vec{\nabla}\mu$ is not orthogonal to the $\vec{\omega}$ lines we can choose $\vec{\nabla}\beta^*$ to be orthogonal to the $\vec{\omega}$ lines and not be in the direction of the $\vec{\nabla}\alpha$ lines, that is we choose β^* not to depend only on α. Since both $\vec{\nabla}\beta^*$ and $\vec{\nabla}\alpha$ are orthogonal to $\vec{\omega}$, $\vec{\omega}$ must take the form:

$$\vec{\omega} = A\vec{\nabla}\alpha \times \vec{\nabla}\beta^* \tag{1.30}$$

However, using equation (1.4) we have:

$$\vec{\nabla} \cdot \vec{\omega} = \vec{\nabla} A \cdot (\vec{\nabla}\alpha \times \vec{\nabla}\beta^*) = 0 \tag{1.31}$$

Which implies that A is a function of α, β^*. Now we can define a new comoving function β such that:

$$\beta = \int_0^{\beta^*} A(\alpha, \beta'^*) d\beta'^*, \qquad \frac{d\beta}{dt} = 0 \tag{1.32}$$

In terms of this function we recover the representation given in equation (1.14):

$$\vec{\omega} = \vec{\nabla}\alpha \times \vec{\nabla}\beta \tag{1.33}$$

Hence we have shown how α, β can be constructed for a known $\vec{v}, \rho$. Notice however, that β is defined in a non unique way since one can redefine β for example by performing the following transformation: $\beta \to \beta + f(\alpha)$ in which $f(\alpha)$ is an arbitrary function. The comoving coordinates α, β serve as labels of the vortex lines. Moreover the vortical flux can be calculated as:

$$\Phi = \int \vec{\omega} \cdot d\vec{S} = \int d\alpha d\beta \tag{1.34}$$

Finally we can use equation (1.8) to derive the function ν for any point s within the flow :

$$\nu(s) = \int_i^s (\vec{v} - \alpha \vec{\nabla}\beta) \cdot d\vec{r} + \nu(i) \tag{1.35}$$

in which i is any arbitrary point within the flow , the result will not depend on the trajectory taken in the case that ν is single valued. If ν is not single valued on should introduce a cut, which the integration trajectory should not cross.

1.2.4 Stationary fluid dynamics

Stationary flows are a unique phenomena of Eulerian fluid dynamics which has no counter part in Lagrangian fluid dynamics . The stationary flow is defined by the fact that the physical fields $\vec{v}, \rho$ do not depend on the temporal coordinate. This however does not imply that the stationary potentials α, β, ν are all functions of spatial coordinates alone. Moreover, it can be shown that choosing the potentials in such a way will lead to erroneous results in the sense that the stationary equations of motion can not be derived from the Lagrangian density $\hat{\mathcal{L}}$ given in equation (1.19). However, this problem can be amended easily as follows. Let us choose α, ν to depend on the spatial coordinates alone. Let us choose β such that:

$$\beta = \bar{\beta} - t \tag{1.36}$$

in which $\bar{\beta}$ is a function of the spatial coordinates. The Lagrangian density $\hat{\mathcal{L}}$ given in equation (1.19) will take the form:

$$\hat{\mathcal{L}} = \rho \left(\alpha - \varepsilon(\rho) - \frac{1}{2}(\vec{\nabla}\nu + \alpha\vec{\nabla}\beta)^2 \right) \tag{1.37}$$

Varying the Lagrangian $\hat{L} = \int \hat{\mathcal{L}} d^3x$ with respect to α, β, ν, ρ leads to the following equations:

$$\begin{aligned}
&\vec{\nabla} \cdot (\rho \hat{\vec{v}}) = 0 \\
&\rho \hat{\vec{v}} \cdot \vec{\nabla}\alpha = 0 \\
&\rho(\hat{\vec{v}} \cdot \vec{\nabla}\bar{\beta} - 1) = 0 \\
&\alpha = \frac{1}{2}\hat{\vec{v}}^2 + h
\end{aligned} \tag{1.38}$$

α is thus the Bernoulli constant (this was also noticed in [13]). Calculations similar to the ones done in previous subsections will show that those equations lead to the stationary Euler equations:

$$\rho(\hat{\vec{v}} \cdot \vec{\nabla})\hat{\vec{v}} = -\vec{\nabla}p(\rho) \tag{1.39}$$

1.2.5 CFD Analysis

For CFD (Computational Fluid Dynamics) analysis one desires a functional which not only gives the stationary flows as its extremum configuration but also that this configuration will be minimal. Furthermore, in order to use simple algorithms such as the conjugate gradient method to minimize the discrete version of the functional it is required that functional will be bounded. Let us look at the negative of the Lagrangian density :

$$-\hat{\mathcal{L}} = \rho \left(\frac{1}{2}(\vec{\nabla}\nu + \alpha\vec{\nabla}\beta)^2 + \varepsilon(\rho) - \alpha \right) \tag{1.40}$$

it contains two positive contributions $(\frac{1}{2}\vec{v}^2, \varepsilon(\rho))$ and a term which can be either positive or negative $(-\alpha)$. Although the positive terms are quadratic in α it can be shown that for suitable choice of the angle between the β and ν surfaces $-\hat{\mathcal{L}}$ can become arbitrarily negative and hence unbound.

1.3 A simpler variational principle of non-stationary fluid dynamics

Lynden-Bell & Katz [7] have shown that an Eulerian variational principle for non-stationary fluid dynamics can be given in terms of two functions the density ρ and the load λ defined in equation (1.22). However, their velocity was given an implicit definition in terms of a partial differential equation and its variations was constrained to satisfy this equation. In this section we will propose a three function variational principle in which the variations of the functions are not constrained in any way, part of our derivation will overlap the formalism of Lynden-Bell & Katz . The three variables will include the density ρ, the load λ and an additional function to be defined in the next subsection. This variational principle is simpler than the Davydov and Seliger & Whitham variational principle [1, 6] which is given in terms of four functions and is more convenient than the Lynden-Bell & Katz [7] variational principle since the variations are not constrained.

1.3.1 Velocity representation

Consider equation (1.24), since $\vec{\omega}$ is orthogonal to $\vec{\nabla}\lambda$ we can write:

$$\vec{\omega} = \vec{K} \times \vec{\nabla}\lambda \tag{1.41}$$

in which $\vec{K}$ is some arbitrary vector field . However, since $\vec{\nabla} \cdot \vec{\omega} = 0$ it follows that $\vec{K} = \vec{\nabla}\theta$ for some scalar function theta. Hence we can write:

$$\vec{\nabla} \times \vec{v} = \vec{\omega} = \vec{\nabla}\theta \times \vec{\nabla}\lambda \tag{1.42}$$

This will lead to:

$$\vec{v} = \theta\vec{\nabla}\lambda + \vec{\nabla}\nu \tag{1.43}$$

For the time being ν is an arbitrary scalar function, the choice of notation will be justified later. Consider now equation (1.23), inserting into this equation

$\vec{v}$ given in equation (1.43) will result in:

$$\frac{d\lambda}{dt} = \frac{\partial \lambda}{\partial t} + \vec{v} \cdot \vec{\nabla} \lambda = \frac{\partial \lambda}{\partial t} + (\theta \vec{\nabla} \lambda + \vec{\nabla} \nu) \cdot \vec{\nabla} \lambda = 0. \tag{1.44}$$

This can be solved for θ, the solution obtained is:

$$\theta = -\left(\frac{\frac{\partial \lambda}{\partial t} + \vec{\nabla} \lambda \cdot \vec{\nabla} \nu}{|\vec{\nabla} \lambda|^2} \right) \tag{1.45}$$

Inserting the above expression for θ into equation (1.43) will yield:

$$\vec{v} = -\frac{\frac{\partial \lambda}{\partial t}}{|\vec{\nabla} \lambda|} \hat{\lambda} + \vec{\nabla} \nu - \hat{\lambda}(\hat{\lambda} \cdot \vec{\nabla} \nu) \equiv -\frac{\frac{\partial \lambda}{\partial t}}{|\vec{\nabla} \lambda|} \hat{\lambda} + \vec{\nabla}^* \nu \tag{1.46}$$

in which $\hat{\lambda} = \frac{\vec{\nabla} \lambda}{|\vec{\nabla} \lambda|}$ is a unit vector perpendicular to the load surfaces and $\vec{\nabla}^* \nu = \vec{\nabla} \nu - \hat{\lambda}(\hat{\lambda} \cdot \vec{\nabla} \nu)$ is the component of $\vec{\nabla} \nu$ parallel to the load surfaces. Notice that the vector $\vec{v} - \vec{\nabla} \nu$ is orthogonal to the load surfaces and that:

$$|\vec{v} - \vec{\nabla} \nu| = (\vec{v} - \vec{\nabla} \nu) \cdot \hat{\lambda} = \theta |\vec{\nabla} \lambda| \Rightarrow \theta = \frac{(\vec{v} - \vec{\nabla} \nu) \cdot \hat{\lambda}}{|\vec{\nabla} \lambda|} \tag{1.47}$$

Further more by construction the velocity field $\vec{v}$ given by equation (1.46) ensures that the load surfaces are comoving. Let us calculate the circulation along λ surfaces:

$$C(\lambda) = \oint_{\lambda} \vec{v} \cdot d\vec{r} = \oint_{\lambda} \vec{\nabla}^* \nu \cdot d\vec{r} = \oint_{\lambda} \vec{\nabla} \nu \cdot d\vec{r} = [\nu]_{\lambda} \tag{1.48}$$

$[\nu]_{\lambda}$ is the discontinuity of ν across a cut which is introduced on the λ surface. Hence in order that circulation $C(\lambda)$ on the load surfaces (and hence everywhere) will not vanish ν must be multiple-valued. Following Lamb [21, page 180, article 132, equation 1] we write ν in the form:

$$\nu = C(\lambda) \bar{\nu}, \qquad [\bar{\nu}]_{\lambda} = 1 \tag{1.49}$$

in terms of $\bar{\nu}$ the velocity is given as:

$$\vec{v} = -\frac{\frac{\partial \lambda}{\partial t}}{|\vec{\nabla} \lambda|} \hat{\lambda} + C(\lambda) \vec{\nabla}^* \bar{\nu} \tag{1.50}$$

And the explicit dependence of the velocity field $\vec{v}$ on the circulation along the load surfaces $C(\lambda)$ is evident.

1.3.2 The variational principle

Consider the action:

$$\begin{aligned} A &\equiv \int \mathcal{L} d^3x dt \\ \mathcal{L} &\equiv \mathcal{L}_1 + \mathcal{L}_2 \\ \mathcal{L}_1 &\equiv \rho(\frac{1}{2}\vec{v}^2 - \varepsilon(\rho)), \qquad \mathcal{L}_2 \equiv \nu[\frac{\partial \rho}{\partial t} + \vec{\nabla} \cdot (\rho \vec{v})] \end{aligned} \tag{1.51}$$

In which $\vec{v}$ is defined by equation (1.46). ν is not a simple Lagrange multiplier since $\vec{v}$ is dependent on ν through equation (1.46). Taking the variational derivative of $\mathcal{L}$ with respect to ν will yield:

$$\delta_\nu \mathcal{L} = \delta\nu[\frac{\partial \rho}{\partial t} + \vec{\nabla} \cdot (\rho \vec{v})] + \rho \vec{v} \cdot \delta_\nu \vec{v} + \nu \vec{\nabla} \cdot (\rho \delta_\nu \vec{v}) \tag{1.52}$$

This can be rewritten as:

$$\delta_\nu \mathcal{L} = \delta\nu[\frac{\partial \rho}{\partial t} + \vec{\nabla} \cdot (\rho \vec{v})] + \rho(\vec{v} - \vec{\nabla}\nu) \cdot \delta_\nu \vec{v} + \vec{\nabla} \cdot (\rho \nu \delta_\nu \vec{v}) \tag{1.53}$$

Now by virtue of equation (1.46):

$$\delta_\nu \vec{v} = \vec{\nabla}^* \delta\nu \tag{1.54}$$

which is parallel to the load surfaces, while from equation (1.43) we see that $\vec{v} - \vec{\nabla}\nu$ is orthogonal to the load surfaces. Hence, the scalar product of those vectors must be null and we can write:

$$\delta_\nu \mathcal{L} = \delta\nu[\frac{\partial \rho}{\partial t} + \vec{\nabla} \cdot (\rho \vec{v})] + \vec{\nabla} \cdot (\rho \nu \vec{\nabla}^* \delta\nu) \tag{1.55}$$

Thus the action variation can be written as:

$$\begin{aligned} \delta_\nu A &= \int d^3x dt \delta\nu[\frac{\partial \rho}{\partial t} + \vec{\nabla} \cdot (\rho \vec{v})] \\ &+ \oint d\vec{S} \cdot \rho \nu \vec{\nabla}^* \delta\nu \end{aligned} \tag{1.56}$$

This will yield the continuity equation using the standard variational procedure. Notice that the surface should include also the "cut" since the ν

function is in general multi valued. Let us now take the variational derivative with respect to the density ρ, we obtain:

$$\begin{aligned} \delta_\rho A &= \int d^3x dt \delta\rho[\frac{1}{2}\vec{v}^2 - w - \frac{\partial \nu}{\partial t} - \vec{v}\cdot\vec{\nabla}\nu] \\ &+ \oint d\vec{S}\cdot\vec{v}\delta\rho\nu + \int d^3x\nu\delta\rho|_{t_0}^{t_1} \end{aligned} \tag{1.57}$$

Hence provided that $\delta\rho$ vanishes on the boundary of the domain and in initial and final times the following equation must be satisfied:

$$\frac{d\nu}{dt} = \frac{1}{2}\vec{v}^2 - w \tag{1.58}$$

This is the same equation as equation (1.10) and justifies the use of the symbol ν in equation (1.43). Finally we have to calculate the variation of the Lagrangian density with respect to λ this will lead us to the following results:

$$\begin{aligned} \delta_\lambda \mathcal{L} &= \rho\vec{v}\cdot\delta_\lambda\vec{v} + \nu\vec{\nabla}\cdot(\rho\delta_\lambda\vec{v}) = \rho(\vec{v} - \vec{\nabla}\nu)\cdot\delta_\lambda\vec{v} + \vec{\nabla}\cdot(\rho\nu\delta_\lambda\vec{v}) \\ &= \rho\theta|\vec{\nabla}\lambda|(\hat{\lambda}\cdot\delta_\lambda\vec{v}) + \vec{\nabla}\cdot(\rho\nu\delta_\lambda\vec{v}) \end{aligned} \tag{1.59}$$

in equation (1.47) was used. Let us calculate $\delta_\lambda\vec{v}$, after some straightforward manipulations one arrives at the result:

$$\delta_\lambda\vec{v} = -\frac{\hat{\lambda}}{|\vec{\nabla}\lambda|}\left[\frac{\partial(\delta\lambda)}{\partial t} + \vec{v}\cdot\vec{\nabla}\delta\lambda\right] + \theta\vec{\nabla}^*\delta\lambda \tag{1.60}$$

Inserting equation (1.60) into equation (1.59) and integrating by parts will yield:

$$\delta_\lambda\mathcal{L} = \delta\lambda\left[\frac{\partial(\rho\theta)}{\partial t} + \vec{\nabla}\cdot(\rho\theta\vec{v})\right] + \vec{\nabla}\cdot[\rho(\delta_\lambda\vec{v}\nu - \theta\vec{v}\delta\lambda)] - \frac{\partial(\rho\theta\delta\lambda)}{\partial t} \tag{1.61}$$

Hence the total variation of the action will become:

$$\begin{aligned} \delta_\lambda A &= \int d^3x dt\delta\lambda\left[\frac{\partial(\rho\theta)}{\partial t} + \vec{\nabla}\cdot(\rho\theta\vec{v})\right] \\ &+ \oint d\vec{S}\cdot(\delta_\lambda\vec{v}\nu - \theta\vec{v}\delta\lambda)\rho - \int d^3x\rho\theta\delta\lambda|_{t_0}^{t_1} \end{aligned} \tag{1.62}$$

Hence choosing $\delta\lambda$ in such a way that the temporal and spatial boundary terms vanish in the above integral will lead to the equation:

$$\frac{\partial(\rho\theta)}{\partial t} + \vec{\nabla} \cdot (\rho\theta\vec{v}) = 0 \quad (1.63)$$

Using the continuity equation (1.1) will lead to the equation:

$$\rho\frac{d\theta}{dt} = 0 \quad (1.64)$$

Hence for $\rho \neq 0$ both λ and θ are comoving. Comparing equation (1.8) to equation (1.43) we see that α is analogue to θ and β is analogue to λ and all those variables are comoving. Furthermore, the ν function in equation (1.43) satisfies the same equation as the ν appearing in equation (1.8) which is equation (1.58). It follows immediately without the need for any additional calculations that $\vec{v}$ given in equation (1.43) satisfies Euler's equations (1.2), the proof for this is given in subsection 1.2.1 in which one should replace α with θ and β with λ. Thus all the equations of fluid dynamics can be derived from the action (1.51) without restricting the variations in any way. The reader should notice an important difference between the current and previous formalism. In the current formalism θ is a dependent variable defined by equation (1.45), while in the previous formalism the analogue quantity α was an independent variational variable. Thus equation (1.64) should be considered as some what complicated second-order partial differential equation (in the temporal coordinate t) for λ which should be solved simultaneously with equation (1.58) and equation (1.1).

1.3.3 Simplified action

The Lagrangian density $\mathcal{L}$ given in equation (1.51) can be written explicitly in terms of the three variational variables ρ, λ, ν as follows:

$$\begin{aligned}
\mathcal{L} &= \hat{\mathcal{L}} + \mathcal{L}_{boundary} \\
\hat{\mathcal{L}} &\equiv \rho\left[\frac{1}{2}\left(\frac{\frac{\partial\lambda}{\partial t} + \vec{\nabla}\lambda\cdot\vec{\nabla}\nu}{|\vec{\nabla}\lambda|}\right)^2 - \frac{1}{2}(\vec{\nabla}\nu)^2 - \frac{\partial\nu}{\partial t} - \varepsilon(\rho)\right] \\
\mathcal{L}_{boundary} &\equiv \frac{\partial(\nu\rho)}{\partial t} + \vec{\nabla}\cdot(\nu\rho\vec{v}) \qquad (1.65)
\end{aligned}$$

Notice that the term $\mathcal{L}_{boundary}$ contains only complete partial derivatives and thus can not contribute to the equations although it can change the boundary conditions. Hence we see that equation (1.1), equation (1.58) and equation (1.64) can be derived using the Lagrangian density $\hat{\mathcal{L}}$ in which $\vec{v}$ is given in terms of equation (1.46) in the relevant equations. Furthermore, after integrating those three equations we can insert the potentials λ, ν into equation (1.46) to obtain the physical velocity $\vec{v}$. Hence, the general barotropic fluid dynamics problem is altered such that instead of solving the four equations (1.1,1.2) we need to solve an alternative set of three equations which can be derived from the Lagrangian density $\hat{\mathcal{L}}$. Notice that the specific choice of the labelling of the λ surfaces is not important in the above Lagrangian density one can replace: $\lambda- > \Lambda(\lambda)$, without changing the Lagrangian functional form. This means that only the shape of the λ surface is important not their labelling. In group theoretic language this implies that the Lagrangian is invariant under an infinite symmetry group and hence should posses an infinite number of constants of motion. In terms of the Lamb type function $\bar{\nu}$ defined in equation (1.49), the Lagrangian density given in equation (1.51) can be rewritten in the form:

$$
\begin{aligned}
\mathcal{L} &= \hat{\mathcal{L}} + \mathcal{L}_{boundary} \\
\hat{\mathcal{L}} &\equiv \rho \left[\frac{1}{2} \left(\frac{\frac{\partial \lambda}{\partial t} + C(\lambda)\vec{\nabla}\lambda \cdot \vec{\nabla}\bar{\nu}}{|\vec{\nabla}\lambda|} \right)^2 - \frac{1}{2}(C(\lambda)\vec{\nabla}\bar{\nu})^2 - C(\lambda)\frac{\partial \bar{\nu}}{\partial t} - \varepsilon(\rho) \right] \\
\mathcal{L}_{boundary} &\equiv \frac{\partial (C(\lambda)\bar{\nu}\rho)}{\partial t} + \vec{\nabla} \cdot (C(\lambda)\bar{\nu}\rho\vec{v})
\end{aligned}
\qquad (1.66)
$$

Which emphasize the dependence of the Lagrangian on the the circulations along the load surfaces $C(\lambda)$ which are given as initial conditions.

1.3.4 Finite element approach for non stationary flows

In the finite element approach we divide the fluid volume into small volumes such that the Lagrangian density can be assumed to be constant in such a small volume:

$$
\int_V \hat{\mathcal{L}} d^3x = \sum_{el} \int_{\Delta V_{el}} \hat{\mathcal{L}} d^3x \simeq \sum_{el} \Delta V_{el} \hat{\mathcal{L}}_{el} \qquad (1.67)
$$

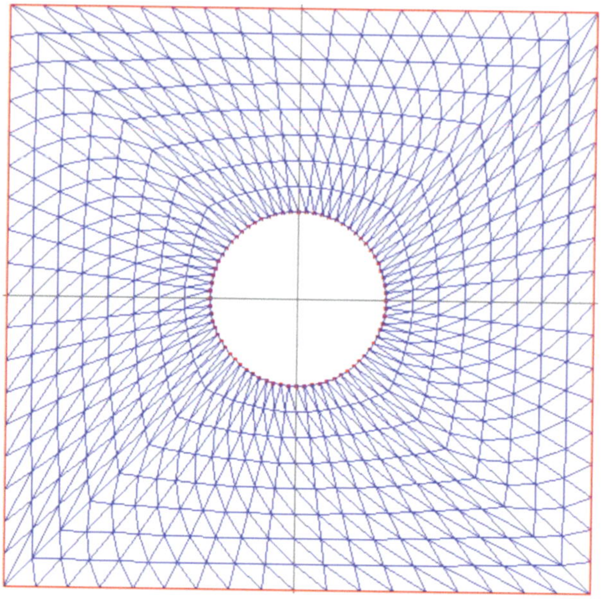

Figure 1.3: Finite element triangles

For example one can dissect a two dimensional flow into small triangle as illustrated in figure (1.3). Using equation (1.65) and equation (1.45) one can estimate $\hat{\mathcal{L}}_{el}$ to be:

$$\hat{\mathcal{L}}_{el} = \rho_{el}\left[\frac{1}{2}\theta^2_{el}(\vec{\nabla}\lambda)^2_{el} - \frac{1}{2}(\vec{\nabla}\nu)^2_{el} - (\frac{\partial\nu}{\partial t})_{el} - \varepsilon(\rho_{el})\right] \qquad (1.68)$$

In which ρ_{el} is assumed to be constant in each element and ν, λ are assumed to be given on the element nodes as illustrated in figure (1.4). In order to calculate the values of ν, λ inside the element one traditionally uses "interpolation functions" ψ_{eli} which are usually low order polynomials. Thus assuming that we are interested in estimating ν, λ in a given element el one can write:

$$\nu \simeq \sum_{i\in el} \nu_{eli}\psi_{eli}, \qquad \lambda \simeq \sum_{i\in el} \lambda_{eli}\psi_{eli} \qquad (1.69)$$

The sum is over all the nodes that belong to a certain element el. Thus we can write the gradient of the above functions as:

$$\vec{\nabla}\nu \simeq \sum_{i\in el} \nu_{eli}\vec{\nabla}\psi_{eli}, \qquad \vec{\nabla}\lambda \simeq \sum_{i\in el} \lambda_{eli}\vec{\nabla}\psi_{eli} \qquad (1.70)$$

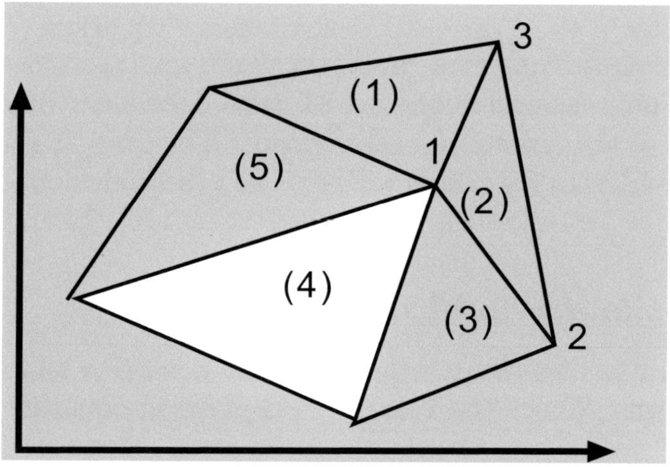

Figure 1.4: Nodes of a finite element triangle

For triangular finite elements $\vec{\nabla}\psi_{eli}$ is constant inside each element, in the more general case it should be evaluated at $\bar{\vec{r}}_{el}$ which is the average value of the nodes coordinates of the element. Hence:

$$(\vec{\nabla}\nu)^2_{el} \simeq \sum_{i,j\in el} \nu_{eli}\nu_{elj}\vec{\nabla}\psi_{eli}\cdot\vec{\nabla}\psi_{elj} \equiv \sum_{i,j\in el} \nu_{eli}\nu_{elj}\; k^{el}_{ij} \tag{1.71}$$

Similarly we obtain:

$$(\vec{\nabla}\lambda)^2_{el} \simeq \sum_{i,j\in el} \lambda_{eli}\lambda_{elj}\; k^{el}_{ij} \tag{1.72}$$

The temporal derivative of ν is evaluated as:

$$(\frac{\partial\nu}{\partial t})_{el} \simeq \sum_{i\in el} \frac{d\nu_{eli}}{dt}\psi_{eli}(\bar{\vec{r}}_{el}) \tag{1.73}$$

Finally we can evaluate θ_{el} using equation (1.45) as follows:

$$\theta_{el} = -\left(\frac{(\frac{\partial\lambda}{\partial t})_{el} + \vec{\nabla}\lambda_{el}\cdot\vec{\nabla}\nu_{el}}{(\vec{\nabla}\lambda)^2_{el}}\right) \tag{1.74}$$

In which $\vec{\nabla}\lambda_{el}$ and $\vec{\nabla}\nu_{el}$ can be estimated using equation (1.70). $(\vec{\nabla}\lambda)^2_{el}$ is given in equation (1.72) and $(\frac{\partial\lambda}{\partial t})_{el}$ is given by:

$$(\frac{\partial\lambda}{\partial t})_{el} \simeq \sum_{i\in el} \frac{d\lambda_{eli}}{dt}\psi_{eli}(\bar{\vec{r}}_{el}) \tag{1.75}$$

The action $A \simeq \int dt \sum_{el} \Delta V_{el} \hat{\mathcal{L}}_{el}$ derived using equation (1.68) is now a functional of N time dependent functions which are the values of ν, λ at the nodes of the finite element mesh and EL time dependent functions ρ which are the values of the density in each element. The 2N+EL Euler -Lagrange equations derived from the action will serve as a finite element approximation to the flow .

1.3.5 Stationary fluid dynamics

For stationary flows we assume that both the density ρ and the load λ are time independent. Hence the velocity field given in equation (1.46) can be written as:

$$\vec{v} = \vec{\nabla}\nu - \hat{\lambda}(\hat{\lambda} \cdot \vec{\nabla}\nu) = \vec{\nabla}^* \nu = C(\lambda)\vec{\nabla}^* \bar{\nu} \tag{1.76}$$

thus the stationary flow is parallel to the load surfaces. From the above equation we see that in the stationary case ν can be written in the form:

$$\nu = \nu_0 - f(\lambda, t) \tag{1.77}$$

in which $f(\lambda, t)$ is an arbitrary function and ν_0 is independent of the temporal coordinate. Hence we can rewrite the velocity $\vec{v}$ as:

$$\vec{v} = \vec{\nabla}^* \nu_0 = C(\lambda)\vec{\nabla}^* \bar{\nu}_0 \tag{1.78}$$

Inserting equation (1.77) and equation (1.78) into equation (1.58) will yield:

$$\frac{\partial f(\lambda, t)}{\partial t} = \frac{1}{2}\vec{v}^2 + w = B(\lambda) \tag{1.79}$$

in which $B(\lambda)$ is the Bernoulli constant. Integrating we obtain:

$$f(\lambda, t) = B(\lambda)t + g(\lambda) \tag{1.80}$$

the arbitrary $g(\lambda)$ function can be absorbed into ν_0 and thus we rewrite equation (1.77) in the form:

$$\nu = \nu_0 - B(\lambda)t \tag{1.81}$$

Further more we can rewrite the conserved quantity θ given in equation (1.45) as:

$$\theta = -\left(\frac{\vec{\nabla}\lambda \cdot \vec{\nabla}\nu}{|\vec{\nabla}\lambda|^2}\right) = -\left(\frac{\hat{\lambda} \cdot \vec{\nabla}\nu_0}{|\vec{\nabla}\lambda|}\right) + t\frac{dB(\lambda)}{d\lambda} \tag{1.82}$$

The Lagrangian density $\mathcal{L}$ given in equation (1.51) can be written in the stationary case taking into account equation (1.78) and equation (1.81) as follows:

$$\begin{aligned}\hat{\mathcal{L}} &= \rho\left[\frac{1}{2}(\hat{\lambda}\cdot\vec{\nabla}\nu)^2 - \frac{1}{2}(\vec{\nabla}\nu)^2 + B(\lambda) - \varepsilon(\rho)\right] \\ &= \rho\left[B(\lambda) - \frac{1}{2}(\vec{\nabla}^*\nu_0)^2 - \varepsilon(\rho)\right] \\ &= \rho\left[B(\lambda) - \frac{1}{2}(C(\lambda)\vec{\nabla}^*\bar{\nu}_0)^2 - \varepsilon(\rho)\right] \qquad (1.83)\end{aligned}$$

Taking the variational derivative of the Lagrangian density $\hat{\mathcal{L}}$ with respect to the mass density ρ will yield the Bernoulli equation (1.79). The variation of the Lagrangian $\hat{L} = \int d^3x\hat{\mathcal{L}}$ with respect to ν_0 will yield the mass conservation equation:

$$\vec{\nabla}^* \cdot (\rho\vec{v}) = 0 \qquad (1.84)$$

this form is equivalent to the standard stationary continuity equation $\vec{\nabla}\cdot(\rho\vec{v}) = 0$ since there is no mass flux orthogonal to the load surfaces. Finally taking the variation of $\hat{L}$ with respect to λ will yield:

$$\rho\left[\frac{dB}{d\lambda} - \vec{v}\cdot\vec{\nabla}^*\left(\frac{\hat{\lambda}\cdot\vec{\nabla}\nu_0}{|\vec{\nabla}\lambda|}\right)\right] = 0 \qquad (1.85)$$

Which can be also obtained by inserting equation (1.82) into equation (1.64). Hence we obtained three equations (1.79,1.84,1.85) for the three spatial functions ρ, λ and ν_0. Admittedly those equations do not have a particularly simple form, we will obtain a somewhat better set of equations in a following section.

1.3.6 CFD analysis

For CFD analysis one desires a functional which not only gives the stationary flows as its extremum configuration but also that this configuration will be minimal. Furthermore, in order to use simple algorithms such as the conjugate gradient method to minimize the discrete version of the functional it is required that functional will be bounded. Let us look at the negative of the Lagrangian density :

$$-\hat{\mathcal{L}} = \rho\left[\frac{1}{2}(\vec{\nabla}^*\nu_0)^2 + \varepsilon(\rho) - B(\lambda)\right] \qquad (1.86)$$

it contains two positive contributions $(\frac{1}{2}\vec{v}^2, \varepsilon(\rho))$ and a term which must be negative $(-B(\lambda))$. Hence by a suitable choice of λ, $-\hat{\mathcal{L}}$ becomes arbitrarily negative and hence unbounded.

1.4 Topological constants of motion

Barotropic fluid dynamics is known to have the helicity topological constant of motion;

$$\mathcal{H} \equiv \int \vec{\omega} \cdot \vec{v} d^3x, \tag{1.87}$$

which is known to measure the degree of knottiness of lines of the vorticity field $\vec{\omega}$ [22].

1.4.1 Representation in terms of the fluid dynamical potentials

Let us write the topological constants given in equation (1.87) in terms of the fluid dynamical potentials α, β, ν, ρ introduced in previous sections, the scalar product $\vec{\omega} \cdot \vec{v}$:

$$\vec{\omega} \cdot \vec{v} = (\vec{\nabla}\alpha \times \vec{\nabla}\beta) \cdot \vec{\nabla}\nu. \tag{1.88}$$

However, since we have the local vector basis: $(\vec{\nabla}\alpha, \vec{\nabla}\beta, \vec{\nabla}\mu)$ we can write $\vec{\nabla}\nu$ as:

$$\vec{\nabla}\nu = \frac{\partial \nu}{\partial \alpha}\vec{\nabla}\alpha + \frac{\partial \nu}{\partial \beta}\vec{\nabla}\beta + \frac{\partial \nu}{\partial \mu}\vec{\nabla}\mu. \tag{1.89}$$

Hence we can write:

$$\vec{\omega} \cdot \vec{v} = \frac{\partial \nu}{\partial \mu}(\vec{\nabla}\alpha \times \vec{\nabla}\beta) \cdot \vec{\nabla}\mu = \frac{\partial \nu}{\partial \mu}\frac{\partial(\alpha, \beta, \mu)}{\partial(x, y, z)}. \tag{1.90}$$

Now we can insert equation (1.90) into equation (1.87) to obtain the expression:

$$\mathcal{H} = \int \frac{\partial \nu}{\partial \mu} d\mu d\alpha d\beta. \tag{1.91}$$

The reader should notice that in some scenarios it may be that the flow domain should be divided into patches in which different definitions of μ, α, β apply to different domains, we do not see this as a limitation for our formalism since the topology of the flow is conserved by the flow equations. In those cases $\mathcal{H}$ should be calculated as sum of the contributions from each patch. We can think about the fluid domain as composed of thin closed tubes of vortex lines each labelled by (α, β). Performing the integration along such a thin tube in the metage direction results in:

$$\oint_{\alpha,\beta} \frac{\partial \nu}{\partial \mu} d\mu = [\nu]_{\alpha,\beta}, \tag{1.92}$$

in which $[\nu]_{\alpha,\beta}$ is the discontinuity of the function ν along its cut. Thus a thin tube of vortex lines in which ν is single valued does not contribute to the helicity integral. Inserting equation (1.92) into equation (1.91) will result in:

$$\mathcal{H} = \int [\nu]_{\alpha,\beta} d\alpha d\beta = \int [\nu] d\Phi, \tag{1.93}$$

in which we have used equation (1.34). Hence:

$$[\nu] = \frac{d\mathcal{H}}{d\Phi}, \tag{1.94}$$

the discontinuity of ν is thus the density of helicity per unit of vortex flux in a tube . We deduce that the Clebsch representation does not entail zero helicity , rather it is perfectly consistent with non zero helicity as was demonstrated above. Further more according to equation (1.10)

$$\frac{d[\nu]}{dt} = 0. \tag{1.95}$$

We conclude that not only is the helicity conserved as an integral quantity of the entire flow domain but also the (local) density of helicity per unit of vortex flux is a conserved quantity as well.

1.4.2 Cross Helicity Conservation via the Noether Theorem

The conservation of helicity $\mathcal{H}$ in ideal barotropic fluid when certain conditions are satisfied in particular, when $\vec{\omega} \cdot \vec{n} = 0$ on the (Lagrangian) surface bounding the volume of integration was discovered by Moffat [22]. Moreau [23] has discussed the conservation of helicity from the group theoretical point of view. In his paper he used an enlarged Arnold symmetry group [24] of fluid element labelling to generate the conservation of helicity . Yahalom

[25, 12] has shown that the symmetry group generating conservation of helicity becomes a very simple one parameter translation group in the space of labels (alpha space) when represented by Lynden-Bell and Katz [7] labelling. The derivation will not be repeated here, the interested reader is referred to the original literature.

1.5 Simplified variational principle for stationary fluid dynamics

In the previous sections we have shown that fluid dynamics can be described in terms of four first order differential equations and in term of an action principle from which those equations can be derived. An alternative derivation in terms of three differential equations one of which (equation (1.64)) is second order has been introduced as well. Those formalisms were shown to apply to both stationary and non-stationary fluid dynamics . In the following a different three functions formalism for stationary fluid dynamics is introduced.

Consider equation (1.13), for a stationary flow it takes the form:

$$\vec{v} \cdot \vec{\nabla} \alpha = 0 \tag{1.96}$$

Hence $\vec{v}$ can take the form:

$$\vec{v} = \frac{\vec{\nabla} \alpha \times \vec{K}}{\rho} \tag{1.97}$$

However, since the velocity field must satisfy the stationary mass conservation equation equation (1.1):

$$\vec{\nabla} \cdot (\rho \vec{v}) = 0 \tag{1.98}$$

We see that $\vec{K}$ must have the form $\vec{K} = \vec{\nabla} N$, where N is an arbitrary function. Thus, $\vec{v}$ takes the form:

$$\vec{v} = \frac{\vec{\nabla} \alpha \times \vec{\nabla} N}{\rho} \tag{1.99}$$

Let us now calculate $\vec{v} \times \vec{\omega}$ in which $\vec{\omega}$ is given by equation (1.14), hence:

$$\begin{aligned}\vec{v} \times \vec{\omega} &= (\frac{\vec{\nabla}\alpha \times \vec{\nabla}N}{\rho}) \times (\vec{\nabla}\alpha \times \vec{\nabla}\beta) \\ &= \frac{1}{\rho}\vec{\nabla}\alpha(\vec{\nabla}\alpha \times \vec{\nabla}N) \cdot \vec{\nabla}\beta \end{aligned} \tag{1.100}$$

Now since the flow is stationary N can be at most a function of the three comoving coordinates $\alpha, \bar{\beta}, \mu$ defined in subsections 1.2.2 and 1.2.4, hence:

$$\vec{\nabla}N = \frac{\partial N}{\partial \alpha}\vec{\nabla}\alpha + \frac{\partial N}{\partial \bar{\beta}}\vec{\nabla}\bar{\beta} + \frac{\partial N}{\partial \mu}\vec{\nabla}\mu \tag{1.101}$$

Inserting equation (1.101) into equation (1.100) will yield:

$$\vec{v} \times \vec{\omega} = \frac{1}{\rho}\vec{\nabla}\alpha\frac{\partial N}{\partial \mu}(\vec{\nabla}\alpha \times \vec{\nabla}\mu) \cdot \vec{\nabla}\bar{\beta} \tag{1.102}$$

Rearranging terms and using vorticity formula (1.14) we can simplify the above equation and obtain:

$$\vec{v} \times \vec{\omega} = -\frac{1}{\rho}\vec{\nabla}\alpha\frac{\partial N}{\partial \mu}(\vec{\nabla}\mu \cdot \vec{\omega}) \tag{1.103}$$

However, using equation (1.27) this will simplify to the form:

$$\vec{v} \times \vec{\omega} = -\vec{\nabla}\alpha\frac{\partial N}{\partial \mu} \tag{1.104}$$

Now let us consider equation (1.3), for stationary flows this will take the form:

$$\vec{\nabla} \times (\vec{v} \times \vec{\omega}) = 0 \tag{1.105}$$

Inserting equation (1.104) into equation (1.105) will lead to the equation:

$$\vec{\nabla}(\frac{\partial N}{\partial \mu}) \times \vec{\nabla}\alpha = 0 \tag{1.106}$$

However, since N is at most a function of $\alpha, \bar{\beta}, \mu$. It follows that $\frac{\partial N}{\partial \mu}$ is some function of α:

$$\frac{\partial N}{\partial \mu} = -F(\alpha) \tag{1.107}$$

This can be easily integrated to yield:

$$N = -\mu F(\alpha) + G(\alpha, \bar{\beta}) \tag{1.108}$$

Inserting this back into equation (1.99) will yield:

$$\vec{v} = \frac{\vec{\nabla}\alpha \times (-F(\alpha)\vec{\nabla}\mu + \frac{\partial G}{\partial \bar{\beta}}\vec{\nabla}\bar{\beta})}{\rho} \tag{1.109}$$

Let us now replace the set of variables $\alpha, \bar{\beta}$ with a new set $\alpha', \bar{\beta}'$ such that:

$$\alpha' = \int F(\alpha) d\alpha, \qquad \bar{\beta}' = \frac{\bar{\beta}}{F(\alpha)} \tag{1.110}$$

This will not have any effect on the vorticity presentation given in equation (1.14) since:

$$\vec{\omega} = \vec{\nabla}\alpha \times \vec{\nabla}\beta = \vec{\nabla}\alpha \times \vec{\nabla}\bar{\beta} = \vec{\nabla}\alpha' \times \vec{\nabla}\bar{\beta}' \tag{1.111}$$

However, the velocity will have a simpler presentation and will take the form:

$$\vec{v} = \frac{\vec{\nabla}\alpha' \times \vec{\nabla}(-\mu + G'(\alpha', \bar{\beta}'))}{\rho} \tag{1.112}$$

in which $G' = \frac{G}{F}$. At this point one should remember that μ was defined in equation (1.25) up to an arbitrary constant which can very between vortex lines. Since the lines are labelled by their $\alpha', \bar{\beta}'$ values it follows that we can add an arbitrary function of $\alpha', \bar{\beta}'$ to μ without effecting its properties. Hence we can define a new μ' such that:

$$\mu' = \mu - G'(\alpha', \bar{\beta}') \tag{1.113}$$

Inserting equation (1.113) into equation (1.112) will lead to a simplified equation for $\vec{v}$:

$$\vec{v} = \frac{\vec{\nabla}\mu' \times \vec{\nabla}\alpha'}{\rho} \tag{1.114}$$

In the following the primes on $\alpha, \bar{\beta}, \mu$ will be ignored. It is obvious that $\vec{v}$ satisfies the following set of equations:

$$\vec{v} \cdot \vec{\nabla}\mu = 0, \qquad \vec{v} \cdot \vec{\nabla}\alpha = 0, \qquad \vec{v} \cdot \vec{\nabla}\bar{\beta} = 1 \tag{1.115}$$

to derive the right hand equation we have used both equation (1.27) and equation (1.14). Hence μ, α are both comoving and stationary . As for $\bar{\beta}$ it satisfies equation (1.38).

By vector multiplying $\vec{v}$ and $\vec{\omega}$ and using equations (1.114,1.14) we obtain:

$$\vec{v} \times \vec{\omega} = \vec{\nabla}\alpha \tag{1.116}$$

this means that both $\vec{v}$ and $\vec{\omega}$ lie on α surfaces and provide a vector basis for this two dimensional surface.

1.5.1 The action principle

In the previous subsection we have shown that if the velocity field $\vec{v}$ is given by equation (1.114) than equation (1.1) is satisfied automatically for stationary flows . To complete the set of equations we will show how the Euler equations (1.2) can be derived from the Lagrangian :

$$\begin{aligned} L &\equiv \int \mathcal{L} d^3x \\ \mathcal{L} &\equiv \rho(\frac{1}{2}\vec{v}^2 - \varepsilon(\rho)) \end{aligned} \tag{1.117}$$

In which $\vec{v}$ is given by equation (1.114) and the density ρ is given by equation (1.27):

$$\rho = \vec{\nabla}\mu \cdot \vec{\omega} = \vec{\nabla}\mu \cdot (\vec{\nabla}\alpha \times \vec{\nabla}\beta) = \frac{\partial(\alpha, \beta, \mu)}{\partial(x, y, z)} \tag{1.118}$$

In this case the Lagrangian density of equation (1.117) will take the form:

$$\mathcal{L} = \rho \left(\frac{1}{2} (\frac{\vec{\nabla}\mu \times \vec{\nabla}\alpha}{\rho})^2 - \varepsilon(\rho) \right) \tag{1.119}$$

and can be seen explicitly to depend on only three functions. The variational derivative of L given in equation (1.117) is:

$$\begin{aligned} \delta L &= \int \delta \mathcal{L} d^3x \\ \delta \mathcal{L} &= \delta\rho(\vec{v}^2 - w(\rho)) + \rho\vec{v} \cdot \delta\vec{v} \end{aligned} \tag{1.120}$$

Let us make arbitrary small variations $\delta\alpha_i = (\delta\alpha, \delta\beta, \delta\mu)$ of the functions $\alpha_i = (\alpha, \beta, \mu)$. Let us define the vector:

$$\vec{\xi} \equiv -\frac{\partial \vec{r}}{\partial \alpha_i}\delta\alpha_i \tag{1.121}$$

This will lead to the equation:

$$\delta\alpha_i = -\vec{\nabla}\alpha_i \cdot \vec{\xi} \tag{1.122}$$

Making a variation of ρ given in equation (1.118) with respect to α_i will yield:

$$\delta\rho = -\vec{\nabla}\cdot(\rho\vec{\xi}) \tag{1.123}$$

(for a proof see for example [10]). Calculating $\delta\vec{v}$ by varying equation (1.114) will give:

$$\delta\vec{v} = -\frac{\delta\rho}{\rho}\vec{v} + \frac{1}{\rho}\vec{\nabla}\times(\rho\vec{\xi}\times\vec{v}) \tag{1.124}$$

Inserting equations (1.123,1.124) into equation (1.120) will yield:

$$\begin{aligned} \delta\mathcal{L} &= \vec{v}\cdot\vec{\nabla}\times(\rho\vec{\xi}\times\vec{v}) - \delta\rho(\frac{1}{2}\vec{v}^2 + w) \\ &= \vec{v}\cdot\vec{\nabla}\times(\rho\vec{\xi}\times\vec{v}) + \vec{\nabla}\cdot(\rho\vec{\xi})(\frac{1}{2}\vec{v}^2 + w) \end{aligned} \tag{1.125}$$

Using the well known vector identity:

$$\vec{A}\cdot\vec{\nabla}\times(\vec{C}\times\vec{A}) = \vec{\nabla}\cdot((\vec{C}\times\vec{A})\times\vec{A}) + (\vec{C}\times\vec{A})\cdot\vec{\nabla}\times\vec{A} \tag{1.126}$$

and the theorem of Gauss we can write now equation (1.120) in the form:

$$\begin{aligned} \delta L &= \oint d\vec{S}\cdot[(\vec{\xi}\times\vec{v})\times\vec{v} + (\frac{1}{2}\vec{v}^2 + w)\vec{\xi}]\rho \\ &+ \int d^3x\vec{\xi}\cdot[\vec{v}\times\vec{\omega} - \vec{\nabla}(\frac{1}{2}\vec{v}^2 + w)]\rho \end{aligned} \tag{1.127}$$

Suppose now that $\delta L = 0$ for a $\vec{\xi}$ such that the boundary term in the above equation is null but that $\vec{\xi}$ is otherwise arbitrary, then it entails the equation:

$$\rho\vec{v}\times\vec{\omega} - \rho\vec{\nabla}(\frac{1}{2}\vec{v}^2 + w) = 0 \tag{1.128}$$

Using the vector identity :

$$\frac{1}{2}\vec{\nabla}(\vec{v}^2) = (\vec{v}\cdot\vec{\nabla})\vec{v} + \vec{v}\times(\vec{\nabla}\times\vec{v}) \tag{1.129}$$

and rearranging terms we recover the stationary Euler equations:

$$\rho(\vec{v}\cdot\vec{\nabla})\vec{v} = -\vec{\nabla}p \tag{1.130}$$

1.5.2 The apparent integration of the stationary Euler equations

Let us now combine equation (1.128) with equation (1.116), this will yield:

$$\vec{\nabla}\alpha = \vec{\nabla}(\frac{1}{2}\vec{v}^2 + w) \tag{1.131}$$

which can easily be integrated to yield:

$$\alpha = \frac{1}{2}\vec{v}^2 + w(\rho) \tag{1.132}$$

in which an arbitrary constant is absorbed into α, this is consistent with equation (1.38) and implies that α is the Bernoulli constant. Let us now insert $\vec{v}$ given by equation (1.114) into the above equation:

$$\alpha = \frac{1}{2}\left(\frac{\vec{\nabla}\mu \times \vec{\nabla}\alpha}{\rho}\right)^2 + w(\rho) \tag{1.133}$$

This can be rewritten as:

$$w(\rho)\rho^2 - \rho^2\alpha + \frac{1}{2}(\vec{\nabla}\mu \times \vec{\nabla}\alpha)^2 = 0 \tag{1.134}$$

By defining:

$$\begin{aligned} \Pi &\equiv \rho^2 \\ W(\Pi) &\equiv w(\rho) \end{aligned} \tag{1.135}$$

We obtain:

$$W(\Pi)\Pi - \Pi\alpha + \frac{1}{2}(\vec{\nabla}\mu \times \vec{\nabla}\alpha)^2 = 0 \tag{1.136}$$

If one chooses an α and μ functions such that the above nonlinear equation has a positive (or zero) solution for Π at every point in space than one obtains a solution of both the Euler and continuity equations in which $\vec{v}$ is given by equation (1.114) and $\rho = +\sqrt{\Pi}$. One should notice, however, that α and μ are not arbitrary functions. α is a function of the load defined in equation (1.22) and thus according to equation (1.24) must satisfy the equation:

$$\vec{\nabla}\alpha \cdot \vec{\omega} = 0 \tag{1.137}$$

that is both the velocity and vorticity fields must lie on the alpha surfaces. While for the velocity $\vec{v}$ this is assured by equation (1.114) this is not assured for the vorticity . Furthermore, μ is not an arbitrary function it is a metage function an thus must satisfy equation (1.27). Moreover, obtaining a solution does not imply that the obtained solution is stable.

1.5.3 CFD Analysis

For CFD analysis one desires a functional which not only gives the stationary flows as its extremum configuration but also that this configuration will be minimal. Furthermore, in order to use simple algorithms such as the conjugate gradient method to minimize the discrete version of the functional it is required that functional will be bounded. Let us look at the Lagrangian density :

$$\mathcal{L} = \rho \left(\frac{1}{2} (\frac{\vec{\nabla}\mu \times \vec{\nabla}\alpha}{\rho})^2 - \varepsilon(\rho) \right) \tag{1.138}$$

it contains a positive contribution $(\frac{1}{2}\vec{v}^2)$ and a term which must be negative $(\varepsilon(\rho))$. However, we can see that for small ρ the first term clearly dominates if we choose $\lim_{\rho\to 0} \varepsilon(\rho) = 0$. Furthermore, if we choose: $\lim_{\rho\to\infty} \varepsilon(\rho) = 0$ the Lagrangian density will be bounded from below and thus can be a suitable functional for CFD analysis.

1.6 Conclusion

In this chapter we have reviewed Eulerian variational principles for barotropic fluid dynamics and introduced a simpler three independent functions variational formalisms for stationary and non-stationary barotropic flows . This is less than the four variables which appear in the standard equations of fluid dynamics which are the velocity field $\vec{v}$ and the density ρ. Some discussion was dedicated to flows with non trivial topology taking into account helicity . This was shown to be related to the non single valuedness of ν. Moreover, it was shown that the discontinuity of ν ($[\nu]$) is a new topological constant of motion which is equal to the amount of helicity per flux.

The above formulation is an advancement in the field theory of barotropic fluids in the sense that was introduced in the preface. That is we have managed to describe the physical phenomena of barotropic flows with less variables than were needed before.

We have discussed numerical aspects of the utilization of three Lagrangians for CFD analysis. Only one of them is bound and could be utilized as a basis for a CFD analysis using the conjugate gradient method. For the other Lagrangians we will probably need to add additional constants of motion constraints to the action as was done by[26, 27] see also [11].

The problem of stability analysis using the described variational principles exceed the scope of this chapter. We suspect that for achieving this we will need to add additional constants of motion constraints to the action as was done by [26, 27] see also [11], hopefully this will be discussed in a future paper.

Bibliography

[1] B. Davydov, "Variational principle and canonical equations for an ideal fluid," *Doklady Akad. Nauk*, vol. 69, 165-168, 1949. (in Russian)

[2] C. Eckart, "Variation Principles of Hydrodynamics ," *Phys. Fluids*, vol. 3, 421, 1960.

[3] J.W. Herivel, "The derivation of the equations of motion of an ideal fluid by Hamiltons principle," *Proc. Camb. Phil. Soc.*, vol. 51, 344, 1955.

[4] J. Serrin, "Mathematical Principles of Classical Fluid Mechanics," In: *Handbuch der Physik*, vol. 8, 148, 1959.

[5] C.C. Lin, "Liquid Helium," In: *Proceedings of the Enrico Fermi International School of Physics Course XXI*, 1963; pp. 120.

[6] R.L. Seliger, and G.B. Whitham, "Variational Principles in Continuum Mechanics," *Proc. Roy. Soc. London A*, vol. 305, 1, 1968.

[7] D. Lynden-Bell, and J. Katz, "Isocirculational Flows and their Lagrangian and Energy principles," *Proceedings of the Royal Society of London. Series A, Mathematical and Physical Sciences*, vol. 378, No. 1773, 179-205, Oct. 8 1981.

[8] J. Katz, and D. Lynden-Bell, "A Lagrangian for Eulerian Fluid Mechanics," *Proc. R. Soc. London A*, vol. 381, 263-274, 1982.

[9] A. Yahalom, and D. Lynden-Bell, "Simplified Variational Principles for Barotropic Fluid Dynamics," Los-Alamos Archives - physics/0603162.

[10] J. Katz, S. Inagaki, and A. Yahalom, "Energy Principles for Self-Gravitating Barotropic Flows : I. General Theory," *Pub. Astro. Soc. Japan*, vol. 45, 421-430, 1993.

[11] A. Yahalom, J. Katz, and S. Inagaki,"Energy Principles for Self-Gravitating Barotropic Flows - Part Two - the Stability of Maclaurin Discs," *Mon. Not. R. Astron. Soc.*, vol. 268, 506-516, 1994.

[12] A. Yahalom, "Energy Principles for Barotropic Flows with Applications to Gaseous Disks", PhD Thesis, The Hebrew University of Jerusalem, Israel, 1996.

[13] A. Yahalom, "Method and System for Numerical Simulation of Fluid Flow ", U.S. Patent 6,516,292, 2003.

[14] A. Yahalom, and G.A. Pinhasi, "Simulating Fluid Dynamics Using a Variational Principle," In: Proceedings of the AIAA Conference, Reno: USA, 2003.

[15] A. Yahalom , G. A. Pinhasi, and M. Kopylenko, "A Numerical Model Based on Variational Principle for Airfoil and Wing Aerodynamics ," In: Proceedings of the AIAA Conference, Reno: USA, 2005.

[16] D. Ophir, A. Yahalom, G.A. Pinhasi, and M. Kopylenko, "A Combined Variational and Multi-grid Approach for Fluid Simulation," In: Proceedings of International Conference on Adaptive Modelling and Simulation (ADMOS), Barcelona:Spain 2005, pp. 295-304.

[17] A. Yahalom "CFD Methods Derived from Simplified Variational Principles," In: Proceedings of AIAA Conference, Reno: USA, 2007.

[18] A. Yahalom "A Finite Element Approach Derived from the Simplified Variational Principle," In: Proceedings of the 9th ASME Engineering Systems Design and Analysis Conference (ESDA), Haifa:Israel 2008.

[19] A. Yahalom, "Stability of Radial Perturbations for non- Uniformly Rotating Self-Gravitating, Finite, Gaseous Disks," *Physics Letters A*, vol. 373, 1170-1176, 2009.
Available: http://dx.doi.org/10.1016/j.physleta.2009.01.060.

[20] D. Lynden-Bell, "Consequence of one spring researching with Chandrasekahar," *Current Science*, vol. 70, No 9., 789-799, 1996.

[21] H. Lamb, Hydrodynamics. Dover Publications, 1945.

[22] H.K. Moffatt, "The degree of knottedness of tangled vortex lines," *J. Fluid Mech.*, vol. 35, 117, 1969.

[23] J.J. Moreau, "Sur les intgrales premires de la dynamique dun fluide parfait barotrope et le thorme de Helmholtz-Kelvin," In: Travaux Sm. Anal. Convexe, 7(1), Montpellier:France 1977, pp. 25.

[24] V.I. Arnold, "Sur un principe variationel pour les ecoulements stationaires," *J. de Mecanique*, vol. 5:1, 29-43, 1966.

[25] A. Yahalom, "Helicity Conservation via the Noether Theorem," *J. Math. Phys.*, vol. 36, 1324-1327, 1995. Available: Los-Alamos Archives solv-int/9407001.

[26] V.I. Arnold, "A variational principle for three-dimensional steady flows of an ideal fluid," *Appl. Math. Mech.*, vol. 29:5, 846-851, 1965.

[27] V.I. Arnold, "On the conditions of nonlinear stability of planar curvilinear flows of an ideal fluid ," *Dokl. Acad. Nauk SSSR*, vol. 162:5, 975-978, 1965.

CHAPTER 2

Advances in the field theory of dissipative electromagnetic fields

Abstract

Equations of motion and energy-momentum tensors are obtained for a dissipative medium sustaining electromagnetic polarizations using a field formalism. A previous work has been simplified by reducing the number of independent vector fields interacting with the sink modes. A relativistic formalism of the same is also suggested.

2.1 Introduction

A field theoretical description of electromagnetism inside a material body in which the electromagnetic field losses energy is difficult. This is in sharp contrast to the situation regarding electromagnetism in vacuum (or in a non dissipative material) in which the Lagrangian is well known.

Any naive attempt will immediately encounter a contradiction with Noether's theorem which dictates that energy of a Lagrangian system without explicit time dependent terms must be conserved and dissipation is prohibited. Hence the mere formulation of such a Lagrangian is an advancement in the classical field theory of electromagnetism in the sense that the theory will encompass a wider range of phenomena that was available previously.

Asher Yahalom (Ed)

Moreover, there are some phenomenological treatments that are not obviously expressible by a Lagrangian formalism and have led to ambiguities and controversies. An instance of this is the dynamics of polarization in dispersive-dissipative materials especially those which exhibit a Quantum resonance type of behaviour. While the equations of motion (shown below in equation (2.10) and equation (2.11)) have been in frequent use, the derivation of these equations from a Lagrangian has been wanting for some time. As a consequence, basic quantities like energy and momentum densities, that are integral parts of the Lagrangian method, have become variously formulated and subject to dispute. A brief account of the history of energy densities is found in [1, 2]. Following that account and adopting a Lagrangian formulation, that work has unambiguously derived an expression for the energy density.

Relativistic treatments of dissipative electromagnetic media, but not based on a Lagrangian formalism, were given in [3]. A comparison of three different approaches for relativistic treatments of dissipative electromagnetic media was undertaken in a book by Fano, Chu and Adler [4]. In this book the authors compared the approach they endorse with two other approaches, denoted the Minkowski and Amperian approaches. The Minkowskian approach was criticized by the authors because it assumes a linear scalar relation between the displacement field $\vec{D}$ and the electric field $\vec{E}$, and between the magnetic field $\vec{H}$ and the magnetic flux density field $\vec{B}$. The authors argue that this relation is not invariant under a general Lorentz transformation and thus is not objective but subjective and can only be regarded as true in a a particular frame of reference. Furthermore, even according to an observer in rest with respect the electromagnetic medium, the linear scalar relations are not generally true since some magnetic materials have a hysteresis curve and thus $\vec{B}$ and $\vec{H}$ are not linearly related, while other materials possess a tensorial dielectric constant instead of a scalar one. Fano, Chu & Adler also claim that the Lorentz force resulting from the Minkowskian formalism contradicts experimental evidence regarding polarized and magnetized materials . The Amperian approach is criticized in the book on the basis that the resulting Lorentz force contradicts experimental evidence regarding magnetized materials. This chapter do not follow the approach of Fano, Chu & Adler, since it is not based on a variational approach and thus one cannot canonically derive an energy expression. Furthermore, Fano, Chu & Adler do not regard the magnetization and polarization of a material as degrees of freedom subject to their own dynamics; rather, these are treated as given source terms for the the electric and magnetic fields. It should be noted however, that both the energy density obtained by them and that using the Amperian approach are positive.

Other approaches, notably for physical systems containing both viscous flows and electromagnetic fields are discussed by Felderhof [5] and Rosensweig [6]. The first work postulates an energy-momentum tensor (equation (3.12) in [5]); in the current formalism we will aim at deriving the energy momentum tensor from a Lagrangian using the well known canonical formalism (see below). The second work [6] is based on an "engineering" type of approach. Rosensweig has postulated balance equations in a moving frame for linear momentum, angular momentum and energy and has derived "constitutive equations" using the first and second laws of thermodynamics. The energy momentum tensor was not considered in this work, nor a Lagrangian to derive the energy momentum tensor from.

It is evident that the energy density, written as T_0^0, is a component of the energy momentum tensor T_β^α ([10]-[13], Greek indices are four dimensional). This is uniquely derivable from the Lagrangian density $\mathcal{L}(u_\kappa, u_{\kappa,\alpha})$, which is composed of the field variables u_κ and of their derivatives $u_{\kappa,\alpha}$. (A symbol after the comma represents a derivative.) The formal definition of the canonical energy-momentum density (as opposed to symmetrized energy-momentum density [13]) is given by:

$$T_\beta^\alpha = \sum_\kappa \frac{\partial \mathcal{L}}{\partial u_{\kappa,\alpha}} u_{\kappa,\beta} - \mathcal{L}\delta_\beta^\alpha \tag{2.1}$$

In our convention the Greek indices take the value 0 for the time component and the numbers $(1, 2, 3)$ for the remaining, space components. The dot over a quantity will signify a time derivative. The leading statement in this paragraph is correct only for closed systems, in this chapter I shall propose a method to energetically close the physical system of an electromagnetic field in a dissipative medium rendering the above statement correct for this type of system.

The expression for the energy found in [1] was positive. The essential idea there was the introduction of added degrees of freedom (named sinks), which are depositories of the the energy residing in the electromagnetic field and in the polarization of the medium. Physically, the sinks are macroscopic modes of motion in the solid (and less obviously, in its surrounding) different from the polarization modes, but weakly interacting with them. These degrees of freedom have been given microscopic interpretation in the thermodynamic limit, in terms of a set of interacting oscillators with a special interaction matrix [7]. The need to include additional degrees of freedom to treat the irreversible behavior of polarizable systems was noted in a classic text [8]. The proof that any set of differential equations can be embedded in a larger set that can be derived from a variational principle was supplied by Bateman [9].

Our aim in the present work is twofold: first we simplify our previous formulation by reducing the number of vector degrees of freedom. Thus, the formalism presented here contains two vector degrees of freedom less than the previous one. The resulting Lagrangian is considerably simpler without compromising the power of our formalism. Secondly, in later sections, we give a relativistic covariant formalism, making the theory valid also in a moving system of reference. The motivation for this is twofold:

1. It is well known that any physical theory should obey the laws of special relativity and should be invariant under Lorentz transformation - in this paper we construct a theory along the same lines as our previous theory [1], which satisfied these requirements.

2. A previous theory did not contain spatial derivatives of certain quantities; the principle of covariance has helped us put those derivatives in.

Further, we discuss the implications.

2.2 Time independent polarizations

Our first task is to formulate the Lagrangian density of the electromagnetic and polarization and magnetization fields for a material whose polarization are static. We use CGS Gaussian units. Connections between the electric displacement field $\vec{D}$ and the electric field $\vec{E}$ and polarization $\vec{P}$, and, likewise, between the magnetic field $\vec{H}$ and the magnetic induction field $\vec{B}$ and magnetization $\vec{M}$ are then provided by the relations:

$$\vec{D} = \vec{E} + 4\pi\vec{P}. \tag{2.2}$$

$$\vec{H} = \vec{B} - 4\pi\vec{M}. \tag{2.3}$$

The above fields satisfy both the homogeneous Maxwell's equations:

$$curl\vec{E} + \frac{1}{c}\dot{\vec{B}} = 0, \qquad div\vec{B} = 0 \tag{2.4}$$

and the inhomogeneous equations:

$$curl\vec{H} - \frac{1}{c}\dot{\vec{D}} = \frac{4\pi}{c}\vec{J}, \qquad div\vec{D} = 4\pi\rho \tag{2.5}$$

We further make the assumption that both free charges ρ and currents $\vec{J}$ are zero.

It is not possible to derive these equations from a Lagrangian density expressed in terms of the fields. However, this difficulty is traditionally overcome by representing the fields in terms of vector $\vec{A}$ and scalar A^0 potentials, as follows:

$$\vec{E} = -\vec{\nabla}A^0 - \frac{1}{c}\dot{\vec{A}}, \qquad \vec{B} = curl\vec{A} \tag{2.6}$$

Using these definitions one can see that the homogenous equations (2.4) are satisfied automatically. This is not much different than the use of Clebsch potential or the load formalism of the previous chapter.

The inhomogeneous equations (2.5) are the outcome of the functional derivation of the Lagrangian below:

$$\begin{aligned}
\mathcal{L} &= \mathcal{L}_{EB} + \mathcal{L}_{PE} + \mathcal{L}_{MH} \\
\mathcal{L}_{EB} &= \frac{1}{8\pi}[\vec{E}^2 - \vec{B}^2] \\
\mathcal{L}_{PE} &= \vec{P}\cdot\vec{E} \\
\mathcal{L}_{MH} &= \vec{M}\cdot\vec{H}
\end{aligned} \tag{2.7}$$

In forming the derivatives it is understood that all field quantities as given in terms of the potentials of equation (2.6) . The energy density is then obtained from equation (2.1) :

$$T_0^0 = \frac{\partial\mathcal{L}}{\partial\dot{\vec{A}}}\cdot\dot{\vec{A}} - \mathcal{L}. \tag{2.8}$$

After some algebra this yields for the Lagrangian of equation (2.7) the result:

$$T_0^0 = \frac{1}{8\pi}[\vec{E}^2 + \vec{H}^2 + (4\pi\vec{M})^2] \tag{2.9}$$

The last term $\vec{M}^2$ can be usually discarded for static magnetization, since it represents nothing but a time-independent constant. However, we include it for reference in future sections, where the electric polarization and the magnetization will become dynamic degrees of freedom.

2.3 Equations of motion with dissipative terms

The equations that treat the development of the polarization $\vec{P}$ by an applied electric field $\vec{E}$, and the magnetization $\vec{M}$ as induced by an external magnetic field $\vec{H}$ are, as appearing in e.g., [14], the following:

$$\ddot{\vec{P}} + \Gamma_e \dot{\vec{P}} + \omega_r^2 \vec{P} = \omega_p^2 \vec{E} \tag{2.10}$$

$$\ddot{\vec{M}} + \Gamma_h \dot{\vec{M}} + \omega_0^2 \vec{M} = F \omega_0^2 \vec{H} \tag{2.11}$$

with all symbols and notation as defined in [14]. Those equation describe a classical analog to resonant quantum systems. These equations hold for $t \geq 0$. Fourier transformed to the frequency (ω) domain, the equations for the frequency components of the fields are written, after a transposition of terms as

$$[-\omega^2 - i\omega\Gamma_e + \omega_r^2]\vec{P}(\omega) = \omega_p^2 \vec{E}(\omega) + \dot{\vec{P}}(t=0) - (i\omega - \Gamma_e)\vec{P}(t=0) \tag{2.12}$$

and similarly for the magnetization. (Here time-derivatives of the polarization were integrated by parts in the manner: $\int_0^\infty \dot{P}(t) e^{i\omega t} dt = [P(t) e^{i\omega t}]_0^\infty - i\omega \int_0^\infty P(t) e^{i\omega t} dt$ and neglecting the field at infinity, etc. As discussed in [8], there can be a polarization at $t = 0$ due to permanent dipoles.) One can introduce the electric and magnetic susceptibilities, defined by

$$\chi_e(\omega) = \frac{\omega_p^2/\omega_r^2}{1 - (\frac{\omega}{\omega_r})^2 - i\frac{\omega\Gamma_e}{\omega_r^2}} \tag{2.13}$$

$$\chi_m(\omega) = \frac{F}{1 - (\frac{\omega}{\omega_0})^2 - i\frac{\omega\Gamma_h}{\omega_0^2}} \tag{2.14}$$

to write the equations of motion for the frequency components as:

$$\begin{aligned} \vec{P}(\omega) &= \chi_e(\omega)[\vec{E}(\omega) + \frac{\dot{\vec{P}}(0) - (i\omega - \Gamma_e)\vec{P}(0)}{\omega_p^2}] \\ \vec{M}(\omega) &= \chi_m(\omega)[\vec{H}(\omega) + \frac{\dot{\vec{M}}(0) - (i\omega - \Gamma_h)\vec{M}(0)}{F\omega_0^2}] \end{aligned} \quad (2.15)$$

The electric and magnetic susceptibilities are complex functions with related imaginary ($\Im$) and real ($\Re$) parts. When the electric losses are sufficiently low $\Gamma_e << 1$, the electric susceptibility can be approximated in the vicinity of the resonance by:

$$\Im[\chi_e] = \frac{\omega_p^2}{\omega_r \Gamma_e} \frac{1}{(\frac{\omega - \omega_r}{\frac{\Gamma_e}{2}})^2 + 1}, \qquad \Re[\chi_e] = -\frac{\omega - \omega_r}{\frac{\Gamma_e}{2}} \Im[\chi_e] \quad (2.16)$$

Analogous relations can be derived for the real and imaginary parts of the magnetic susceptibility for $\Gamma_h << 1$:

$$\Im[\chi_m] = \frac{F\omega_0}{\Gamma_h} \frac{1}{(\frac{\omega - \omega_0}{\frac{\Gamma_h}{2}})^2 + 1}, \qquad \Re[\chi_m] = -\frac{\omega - \omega_0}{\frac{\Gamma_h}{2}} \Im[\chi_e] \quad (2.17)$$

Returning now to equation (2.10) and equation (2.11) , for self-consistency these have to be supplemented by the Maxwell-equations given in the previous section and derived from the Lagrangian of equation (2.7) that contain the contravariant four-vectors $A^\alpha = (A^0, \vec{A})$.

To take account of the dissipative nature of the processes we introduce a set of new fields, designated canonical fields. These will be formally distinguished from the previous, physical fields by writing them in lower case symbols. The introduction of these fields follows theories by [15] -[17]. The canonical fields (taken to be real) are made to be part of a Lagrangian formulation. We note that for the preceding physical quantities (written in upper case symbols), this cannot be achieved [15].

To make our new notation explicit, we shall work with electric polarization p^i and magnetization m^i and the electromagnetic fields e^i, h^i (Latin indices are three dimensional) and the (contravariant) vector-potentials a^α. It will be later demonstrated that these new, canonical fields contain a time dependent factor which (partly, but not fully) neutralizes the dissipative process. Otherwise, the choice of the fields (and of the various constant factors) is guided by the requirement that we recapture the form of the equations of motion and the energy density currently widely employed in solid state optics in the appropriate limiting case of no dissipation. (This requirement impairs at times the formal symmetry between the modes linked with the polarization and the magnetization, respectively.) The physical meaning of the canonical fields will be made clear later by comparing their equations of motion with those of the physical fields. (For a similar procedure see [15].)

In addition, we shall introduce two new scalar fields, the so called "sink" fields: the first being Ψ, that incorporates macroscopic degrees of freedom associated with the dissipation of the polarization (analogous to the degree of freedom denoted with the same symbol in [17]) and the second, the sink-field Φ which accounts for the dissipation in the magnetic mode (and originates physically in atomic spin or angular momenta in the solid).

The Ψ and Φ functions are representations of the sink modes and thus they collectively represent a huge amount of microscopic degrees of freedom which take away some of the energy of the electromagnetic field. Different macroscopic materials such as air and water have quite a different microscopic structure, nevertheless, we believe that their interaction with the electromagnetic field can be conceptually unified in a model of the type described in our papers, the microscopic difference between materials being represented by different coupling constants. The experimental consequences of the theory are given by equations (2.10) and (2.11), however, the constants appearing in those equations can not be derived in this framework and a deeper microscopic theory is needed. The situation is not very different from viscous fluid dynamics in which both air and water which are completely different microscopic systems are described by the Navier-Stokes equations with different viscosity coefficients.

2.4 The Lagrangian with sink terms

The addition of sink fields requires the extension of the Lagrangian as follows:

$$\mathcal{L} = \mathcal{L}_{eb} + \mathcal{L}_{pe} + \mathcal{L}_{mh} + \mathcal{L}_{p\Psi} + \mathcal{L}_{m\Phi} + \mathcal{L}_{\Psi} + \mathcal{L}_{\Phi} \qquad (2.18)$$

In this Lagrangian we have first the Lagrangian of equation (2.7) , there written in terms of the electromagnetic fields, and now rewritten in the new lower-case variables as $\mathcal{L}_{eb}+\mathcal{L}_{pe}+\mathcal{L}_{mh}$. The above expression also contains the

Lagrangians in the polarization modes, the magnetization modes, the electromagnetic fields, the Ψ and Φ sink fields and then the interaction-Lagrangians between polarization and electric field, magnetization and magnetic field and, finally, the polarization and magnetization with their respective sink fields. The various terms are written out as:

$$\mathcal{L}_{eb} = \frac{1}{8\pi}[e_1^2 - b_1^2] \tag{2.19}$$

$$\mathcal{L}_{pe} = pe_1 \tag{2.20}$$

$$\mathcal{L}_{mh} = mh_1 \tag{2.21}$$

$$\mathcal{L}_{p\Psi} = \frac{1}{2\omega_p^2}(\dot{p}^2 - (\omega_r^2 - \dot{\Psi}^2)p^2 - 2p\dot{p}\dot{\Psi}) \tag{2.22}$$

$$\mathcal{L}_{m\Phi} = \frac{1}{2F\omega_0^2}(\dot{m}^2 - (\omega_0^2 - \dot{\Phi}^2)m^2 - 2m\dot{m}\dot{\Phi}) \tag{2.23}$$

having chosen the x (or 1) axis as the direction of the polarization and magnetization vectors . For $\mathcal{L}_\Psi$ and $\mathcal{L}_\Phi$ see [1].

We next derive $\mathcal{L}$ with respect to each degree of freedom. In accordance with the Euler-Lagrange equations, this procedure leads to the equations of motion. In detail, for the variables p and m the equations of motion take the following form:

$$\ddot{p} - (\ddot{\Psi} + \dot{\Psi}^2 - \omega_r^2)p = \omega_p^2 e_1 \tag{2.24}$$

$$\ddot{m} - (\ddot{\Phi} + \dot{\Phi}^2 - \omega_0^2)m = F\omega_0^2 h_1 \tag{2.25}$$

The electromagnetic equations of motion are identical to Maxwell's equations in the new (small letter) variables. Lastly, for the sink coordinates one finds

$$\ddot{\Psi} + \frac{\rho}{\kappa}\Psi = \frac{1}{\kappa\omega_p^2}\frac{\partial(p(\dot{p} - p\dot{\Psi}))}{\partial t} \tag{2.26}$$

$$\ddot{\Phi} + \frac{\nu}{\lambda}\Phi = \frac{1}{\lambda F\omega_0^2}\frac{\partial(m(\dot{m} - m\dot{\Phi}))}{\partial t} \tag{2.27}$$

The expression for the energy density T_0^0 is found from equation (2.1) , namely,

$$T_0^0 = \sum_\kappa \frac{\partial\mathcal{L}}{\partial\dot{u}_\kappa}\dot{u}_\kappa - \mathcal{L} \tag{2.28}$$

with the summation over all degrees of freedom. We separate the electromagnetic, the polarization and the magnetization parts

$$T^0_0 = (T^0_0)_{e,h} + (T^0_0)_{p,\Psi} + (T^0_0)_{m,\Phi} \tag{2.29}$$

where the first term has the well known form of the electromagnetic energy density and is already given in equation (2.9) .

The others take the forms:

$$(T^0_0)_{p,\Psi} = \frac{1}{2\omega_p^2}\left[(\dot{p} - p\dot{\Psi})^2 + \omega_r^2 p^2\right] + \frac{\kappa}{2}\dot{\Psi}^2 + \frac{\rho}{2}\Psi^2 \tag{2.30}$$

$$= \frac{1}{2\omega_p^2}(\dot{p}^2 + \omega_r^2 p^2) + \frac{\kappa}{2}\dot{\Psi}^2 + \frac{\rho}{2}\Psi^2 + \frac{1}{2\omega_p^2}p^2\dot{\Psi}^2 - \frac{1}{2\omega_p^2}\frac{\partial p^2}{\partial t}\dot{\Psi} \tag{2.31}$$

and

$$(T^0_0)_{m,\Phi} = \frac{1}{2F\omega_0^2}\left[(\dot{m} - m\dot{\Phi})^2 + \omega_0^2 m^2\right] + \frac{\lambda}{2}\dot{\Phi}^2 + \frac{\nu}{2}\Phi^2 \tag{2.32}$$

$$= \frac{1}{2F\omega_0^2}(\dot{m}^2 + \omega_0^2 m^2) + \frac{\lambda}{2}\dot{\Phi}^2 + \frac{\nu}{2}\Phi^2 + \frac{1}{2F\omega_0^2}m^2\dot{\Phi}^2 - \frac{1}{2F\omega_0^2}\frac{\partial m^2}{\partial t}\dot{\Phi} \tag{2.33}$$

The first-written forms are sums of squares with non-negative coefficients; this ensures that each part of the energy density is positive (non-negative). The usual expressions for the energy density, e.g. in [13] or [14], differ from the above by the presence of the terms in Φ and Ψ and their time derivatives. In the following section we shall eliminate these variables by making use of the equations of motion, equation (2.26) and equation (2.27) .

2.5 A simple solution

A fully solvable case was described in [1]. The results can be summarized as follows. First we have obtained under assumptions outlined in [1] solutions for $\Psi(t)$ (and analogously for $\Phi(t)$):

$$\begin{aligned}\Psi(t) &= \frac{1}{2}\ln\left[\frac{(1+\frac{p^2(t)}{\kappa\omega_p^2})}{(1+\frac{p^2(0)}{\kappa\omega_p^2})}\right] \\ &+\left[\frac{\Gamma_e}{2}(1+\frac{p^2(0)}{\kappa\omega_p^2}) - \frac{p(0)\dot{p}(0)}{\kappa\omega_p^2}\right]\int_0^t \frac{dt'}{1+\frac{p^2(t')}{\kappa\omega_p^2}} \qquad (2.34)\end{aligned}$$

Those solutions are inserted in equation (2.31) and equation (2.33) to obtain the results:

$$(T_0^0)_{p,\Psi} = \frac{1}{2\omega_p^2}\Big(\frac{\dot{p}^2(t)}{1+\frac{p^2(t)}{\kappa\omega_p^2}} + \omega_r^2 p^2(t)\Big) \qquad (2.35)$$

$$(T_0^0)_{m,\Phi} = \frac{1}{2F\omega_0^2}\Big(\frac{\dot{m}^2(t)}{1+\frac{m^2(t)}{\lambda F\omega_0^2}} + \omega_0^2 m^2(t)\Big) \qquad (2.36)$$

The main results of this section, equation (2.35) and its analogue for the magnetization energy density, are exact and contain non-perturbative corrections to the energy density, due to the presence of the sink degrees of freedom. While exact, they are model dependent in the sense that sinks represented by different Lagrangians would lead to different energy densities. Notice, however, that both of those expressions are positive.

2.6 Interpretation of the fields

The relation of the canonical fields to the physical fields in the presence of dissipation and sink modes was derived in [1]. In [1] we regained the original equations of motion, equation (2.10) and equation (2.11) , for the physical polarization variable by postulating:

$$\begin{aligned} p(t) &= e^{\Psi(t)}P(t) \qquad (2.37)\\ m(t) &= e^{\Phi(t)}M(t) \qquad (2.38)\end{aligned}$$

and

$$\begin{aligned} \vec{e}(t) &= e^{\Psi(t)}\vec{E}(t) \qquad (2.39)\\ \vec{h}(t) &= e^{\Phi(t)}\vec{H}(t). \qquad (2.40)\end{aligned}$$

This provides a physical meaning for *all* the "canonical" variables, as those fields in which the decay has been reinstated. On the other hand, the decay is itself dependent on the fields. (Cf. [20].) Furthermore, the Maxwell equations for the physical fields are also modified in the dissipative-polarizable medium. This feature (of a *modified* Maxwell equation) also appears in [16] and [17] (eq. (14) and eq. (12), respectively).

2.7 A covariant dissipative Lagrangian density

In this section we introduce a covariant Lagrangian density which is a Lorentz invariant generalization of the Lagrangian density described in equation (2.18). Again we introduce a set of new fields, which are formally distinguished from the previous, physical fields by writing them in lower case symbols. Explicitly, we shall work with the scaled polarization and electromagnetic tensors $p^{\alpha\beta}$ and $f^{\alpha\beta}$ as defined by Jackson [13] and other lower case quantities that were introduced earlier. The Greek indices are the usual special relativity four dimensional indices as defined by Jackson [13]. Notice that the Lagrangian introduced here will contain less terms since both the magnetization and the polarization are now part of the polarization tensor $p^{\alpha\beta}$. This Lagrangian has the following parts:

$$\mathcal{L} = \mathcal{L}_{fp} + \mathcal{L}_{p\Psi} + \mathcal{L}_{\Psi} \tag{2.41}$$

In which $\mathcal{L}_{fp}$ is defined as:

$$\begin{aligned} \mathcal{L}_{fp} &= \mathcal{L}_f + \mathcal{L}_{fj} \\ \mathcal{L}_f &= -\frac{1}{16\pi} f_{\alpha\beta} f^{\alpha\beta} \\ \mathcal{L}_{FJ} &= -\frac{1}{c} j_\alpha a^\alpha \qquad j_\alpha = c\partial^\beta p_{\beta\alpha}. \end{aligned} \tag{2.42}$$

$\mathcal{L}_{p\Psi}$ is obtained by generalizing equation (2.22) as follows:

$$\mathcal{L}_{p\Psi} = \frac{1}{2\hat{\omega}_p^2}\left[\partial_\mu p^{\alpha\beta}\partial^\mu p_{\alpha\beta} - (\hat{\omega}_r^2 - \partial_\mu \Psi \partial^\mu \Psi) p_{\alpha\beta} p^{\alpha\beta} - 2 p_{\alpha\beta}\partial_\mu p^{\alpha\beta}\partial^\mu \Psi\right] \tag{2.43}$$

$\mathcal{L}_{p\Psi}$ can be written as a sum of two terms: one depending on the magnetization, while the other depends on the polarization.

$$
\begin{aligned}
\mathcal{L}_{p\Psi} &= \mathcal{L}_{\vec{p}\Psi} + \mathcal{L}_{\vec{m}\Psi} \\
\mathcal{L}_{\vec{p}\Psi} &= \frac{1}{\hat{\omega}_p^2}[-\frac{1}{c^2}(\partial_t \vec{p})^2 + (\partial_i \vec{p})^2 + (\hat{\omega}_r^2 - \frac{1}{c^2}(\partial_t \Psi)^2 + (\vec{\nabla}\Psi)^2)\vec{p}^2 \\
&+ \frac{2}{c^2}\vec{p}\cdot\dot{\vec{p}}\dot{\Psi} - 2\vec{p}\cdot\partial_i \vec{p}\partial_i \Psi] \\
\mathcal{L}_{\vec{m}\Psi} &= -\frac{1}{\hat{\omega}_p^2}[-\frac{1}{c^2}(\partial_t \vec{m})^2 + (\partial_i \vec{m})^2 + (\hat{\omega}_r^2 - \frac{1}{c^2}(\partial_t \Psi)^2 + (\vec{\nabla}\Psi)^2)\vec{m}^2 \\
&+ \frac{2}{c^2}\vec{m}\cdot\dot{\vec{m}}\dot{\Psi} - 2\vec{m}\cdot\partial_i \vec{m}\partial_i \Psi] \qquad (2.44)
\end{aligned}
$$

Finally $\mathcal{L}_\Psi$ is obtained by generalizing an equation from [1] as follows:

$$
\mathcal{L}_\Psi = \frac{1}{2}[\hat{\kappa}\partial_\mu \Psi \partial^\mu \Psi - \rho\Psi^2] = \frac{1}{2}[\hat{\kappa}(\frac{1}{c^2}(\partial_t \Psi)^2 - (\vec{\nabla}\Psi)^2) - \rho\Psi^2] \qquad (2.45)
$$

The field equations with respect to $p^{\alpha\beta}$, Ψ and the electromagnetic fields can be obtained by taking the variational derivative of $\mathcal{L}$ given in equation (2.41) with respect to $p^{\alpha\beta}$, Ψ and a^α. Maxwell's macroscopic field equations can be derived by taking the variational derivative of equation (2.42) with respect to a^α. Taking the variational derivative of $\mathcal{L}$ with respect to Ψ leads to the equation:

$$
\hat{\kappa}\partial_\mu \partial^\mu \Psi + \rho\Psi = \frac{1}{\hat{\omega}_p^2}\partial^\mu[p^{\alpha\beta}(\partial_\mu p_{\alpha\beta} - \partial_\mu \Psi p_{\alpha\beta})] \qquad (2.46)
$$

Taking the variational derivative of $\mathcal{L}$ with respect to $p_{\alpha\beta}$ leads to the equation:

$$
\partial_\mu \partial^\mu p_{\alpha\beta} + (\hat{\omega}_r^2 - \partial_\mu \Psi \partial^\mu \Psi - \partial_\mu \partial^\mu \Psi)p_{\alpha\beta} = -\frac{1}{2}\hat{\omega}_p^2 f_{\alpha\beta} \qquad (2.47)
$$

In the "homogeneous" case in which the spatial derivatives of both the polarization tensor $p^{\alpha\beta}$ and the scalar field Ψ vanish, the tensor equation (2.47) can be written as two vector equations:

$$
\begin{aligned}
\ddot{\vec{p}} - (\ddot{\Psi} + \dot{\Psi}^2 - \omega_r^2)\vec{p} &= \omega_p^2 \vec{e} \qquad (2.48) \\
\ddot{\vec{m}} - (\ddot{\Psi} + \dot{\Psi}^2 - \omega_r^2)\vec{m} &= -\omega_p^2 \vec{b} \qquad (2.49)
\end{aligned}
$$

In which $\omega_r^2 = c^2\hat{\omega}_r^2$ and $\omega_p^2 = \frac{1}{2}c^2\hat{\omega}_p^2$. We can clearly see that equation (2.48) is the same as equation (2.24) . As for equation (2.49) by using the equality

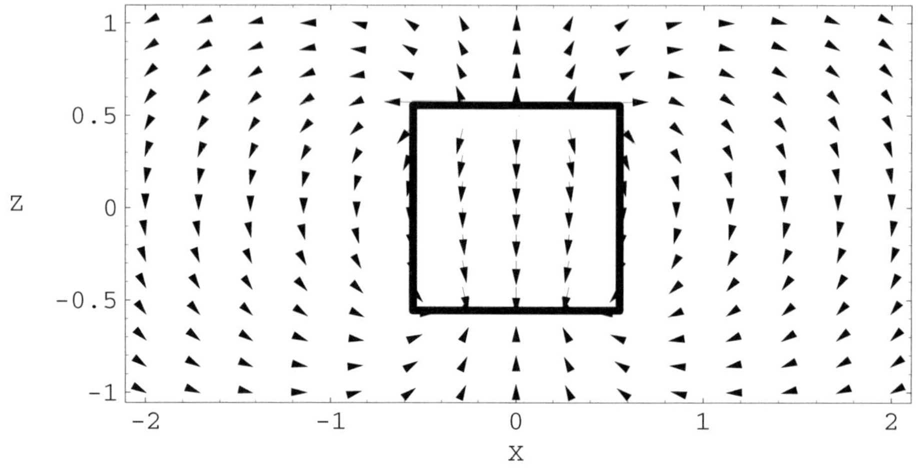

Figure 2.1: Magnetic field lines in a box of permanent uniform magnetization - a cross section view. The direction of magnetization can be inferred from the way the magnetic field lines look **outside** the box magnet. The normal component of the field changes sign across the material boundary, unlike in the dielectric case.

(2.3) this can be rewritten as:

$$\ddot{\vec{m}} - (\ddot{\Psi} + \dot{\Psi}^2 - \omega_0^2)\vec{m} = -\omega_p^2 \vec{h} \tag{2.50}$$

In which $\omega_0^2 = \omega_r^2 + 4\pi\omega_p^2$. This equation can be identified with equation (2.25) by choosing $F = -\frac{\omega_p^2}{\omega_0^2}$ and equating the magnetic and electric dissipative modes ($\Phi \equiv \Psi$). Notice that in magnetized materials the magnetic field is in a direction opposite to the magnetization as can be seen from figure 2.1, which explains the negative sign in the above equation.

2.7.1 The energy momentum tensor of the dissipative Lagrangian density

Using the dissipative Lagrangian density defined in equation (2.41) we can calculate the energy momentum tensor by using the formula given in (2.1). Doing so we obtain the following expression:

$$T_\alpha^\beta = T_\alpha^\beta[fp] + T_\alpha^\beta[p\Psi] + T_\alpha^\beta[\Psi] \tag{2.51}$$

$T_\alpha^\beta[fp]$ is given by:

$$T_\alpha^\beta[fp] = \frac{1}{4\pi}\partial_\alpha a_\mu f^{\mu\beta} + a_\mu \partial_\alpha p^{\mu\beta} - \delta_\alpha^\beta \mathcal{L}_{fp}. \tag{2.52}$$

One would expect that the relativistic energy density in equation (2.52) would smoothly approach the non-relativistic expression in equation (2.9) as $c \to \infty$. Apparently this is not the case, because of the different signs of the quadratic magnetization term. In truth, when one goes deeper, into the atomic theory of magnetization, one recognizes that both the orbital and spin contributions to the magnetization M are proportional to the Bohr magneton $\frac{e\hbar}{2mc}$. This vanishes in a theory where $c \to \infty$. It remains to obtain expressions for $T_\alpha^\beta[p\Psi]$ and $T_\alpha^\beta[\Psi]$; these are given below:

$$\begin{aligned} T_\alpha^\beta[p\Psi] &= \frac{1}{\hat{\omega}_p^2}[\partial^\beta p^{\gamma\theta}\partial_\alpha p_{\gamma\theta} - p^{\gamma\theta}(\partial^\beta\Psi\partial_\alpha p_{\gamma\theta} + \partial_\alpha\Psi\partial^\beta p_{\gamma\theta}) \\ &+ \partial^\beta\Psi\partial_\alpha\Psi p_{\gamma\theta}p^{\gamma\theta}] - \delta_\alpha^\beta\mathcal{L}_{p\Psi} \end{aligned} \quad (2.53)$$

and

$$T_\alpha^\beta[\Psi] = \hat{\kappa}\partial^\beta\Psi\partial_\alpha\Psi - \delta_\alpha^\beta\mathcal{L}_\Psi \quad (2.54)$$

Let us look at the energy density T_0^0. For $T_0^0[\Psi]$ we obtain the positive quantity

$$T_0^0[\Psi] = \frac{1}{2}[\frac{\hat{\kappa}}{c^2}\dot{\Psi}^2 + \hat{\kappa}(\vec{\nabla}\Psi)^2 + \rho\Psi^2] \quad (2.55)$$

For $T_0^0[p\Psi]$ we obtain

$$T_0^0[p\Psi] = \frac{1}{\hat{\omega}_p^2}[(\partial_\mu\vec{m} - \partial_\mu\Psi\vec{m})^2 - (\partial_\mu\vec{p} - \partial_\mu\Psi\vec{p})^2 + \hat{\omega}_r^2(\vec{m}^2 - \vec{p}^2)] \quad (2.56)$$

This expression contains positive magnetization contributions and negative polarization contributions. The positive magnetization contributions can overcome the negative magnetization contribution of $T_0^0[fp]$ given by:

$$T_0^0[fp] = \frac{1}{8\pi}[\vec{e}^2 + \vec{h}^2 - (4\pi\vec{m})^2] + \vec{\nabla}\cdot(\frac{\phi\vec{e}}{4\pi} + \vec{a}\times\vec{m}). \quad (2.57)$$

provided that $\frac{\hat{\omega}_r^2}{\hat{\omega}_p^2} > 2\pi$. However, the negative polarization contribution to T_0^0 results in the uncertainty whether the total polarisation contribution is negative or positive, since this depends on the solution of equations (2.46,2.47) together with the macroscopic Maxwell equations. Those equations are difficult to solve since they are complicated non-linear partial differential equations. In the following subsection we will present a solution for a particular, simple situation. Notice that this situation is completely different from the situation in the non-covariant model for which the positiveness of the energy density was assured. To conclude this subsection we calculate the contributions to T^i_0:

$$T^i_0 = T^i_0[fp] + T^i_0[p\Psi] + T^i_0[\Psi] \qquad (2.58)$$

since $T^i_0[fp]$ given by:

$$T^i_0[fp] = [\frac{1}{4\pi}\vec{e}\times\vec{h} + \frac{1}{c}\partial_t(\vec{m}\times\vec{a} - \frac{\phi\vec{e}}{4\pi}) + \vec{\nabla}\times(\frac{\phi\vec{h}}{4\pi})]^i \qquad (2.59)$$

is the Poynting vector (plus additional boundary terms and total time derivatives). The expression given by equation (2.59) is almost identical to the expression obtained by Jackson ([13] chapter 12) for a free electromagnetic field. Its "complex" form is due to our choice of the canonical energy momentum tensor. If one is to avoid the "complication" one should use the symmetric energy momentum tensor as explained by Jackson ([13] chapter 12). We can consider $T^i_0[p\Psi]$ and $T^i_0[\Psi]$ as corrections to the Poynting vector. Those are given by:

$$\begin{aligned} T^i_0[p\Psi] &= \frac{2}{c\hat{\omega}_p^2}[\vec{\nabla}p^k\dot{p}^k - \vec{\nabla}m^k\dot{m}^k - p^k(\vec{\nabla}\Psi\dot{p}^k + \dot{\Psi}\vec{\nabla}p^k) \\ &+ m^k(\vec{\nabla}\Psi\dot{m}^k + \dot{\Psi}\vec{\nabla}m^k) - \vec{\nabla}\Psi\dot{\Psi}(\vec{m}^2 - \vec{p}^2)] \end{aligned} \qquad (2.60)$$

and

$$T^i_0[\Psi] = -\frac{\hat{\kappa}}{c}\vec{\nabla}\Psi\dot{\Psi} \qquad (2.61)$$

One should notice that both these corrections vanish for the homogeneous case in which the spatial derivatives vanish.

2.7.2 A covariant solvable case

In this final subsection we would like to elucidate the nature of the energy density component T^0_0 by introducing a simple solution of equation (2.46). We assume a homogeneous situation in which all spatial derivative vanish. Furthermore, we assume the $\rho = 0$ and introduce the constant $\kappa = \frac{\hat{\kappa}}{c^2}$. In this case we obtain the equation

$$\kappa\ddot{\Psi} = \frac{1}{\omega_p^2}\partial_t[\vec{m}\cdot\dot{\vec{m}} - \dot{\Psi}\vec{m}^2 - \vec{p}\cdot\dot{\vec{p}} + \dot{\Psi}\vec{p}^2] \qquad (2.62)$$

From this equation we can obtain an expression for $\dot{\Psi}$, as follows:

$$\dot{\Psi} = (\frac{1}{2})\frac{\partial_t(\vec{m}^2 - \vec{p}^2)}{\kappa\omega_p^2 + \vec{m}^2 - \vec{p}^2} \tag{2.63}$$

(In which we have neglected a constant that can be considered small). Using equation (2.63) we can rewrite equation (2.55) as

$$T_0^0[\Psi] = \frac{1}{2}\kappa\dot{\Psi}^2 = \frac{1}{8}\kappa\left(\frac{\partial_t(\vec{m}^2 - \vec{p}^2)}{\kappa\omega_p^2 + \vec{m}^2 - \vec{p}^2}\right)^2 \tag{2.64}$$

To this we need to add the homogeneous energy density

$$T_0^0[p\Psi] = \frac{1}{2\omega_p^2}[(\dot{\vec{m}} - \dot{\Psi}\vec{m})^2 - (\dot{\vec{p}} - \dot{\Psi}\vec{p})^2 + \omega_r^2(\vec{m}^2 - \vec{p}^2)] \tag{2.65}$$

We will consider to separate cases:

1. $\vec{p} = 0, \vec{m} = m\hat{x}$. In this case we obtain

$$T_0^0[\Psi] + T_0^0[p\Psi] = \frac{1}{2\omega_p^2}[\omega_r^2 m^2 + \frac{\dot{m}^2}{1 + \frac{m^2}{\kappa\omega_p^2}}] \tag{2.66}$$

which is very similar to equation (2.36). This expression is positive and with the correct choice of parameters can overcome the negative magnetization contribution of $T_0^0[fp]$

2. $\vec{p} = p\hat{x}, \vec{m} = 0$. In this case we obtain

$$T_0^0[\Psi] + T_0^0[p\Psi] = \frac{1}{2}[\frac{\kappa\dot{p}^2}{p^2 - \kappa\omega_p^2} - \frac{\omega_r^2}{\omega_p^2}p^2] \tag{2.67}$$

which is very different from equation (2.35). This expression is positive only if $p^2 > \kappa\omega_p^2$ and if p changes fast enough to overcome the negative part $-\frac{\omega_r^2}{\omega_p^2}p^2$; otherwise, the energy density is negative.

2.8 Conclusion

In this chapter the Lagrangian method for a dissipative medium (capable also of sustaining electric and/or magnetic polarizations) was carried out to obtain equations of motion, energy-momentum densities, etc. To apply the method it has been found necessary to introduce additional degrees of freedom ("sink-modes"), associated with decay mechanisms in the electrical and magnetic modes.

This is an advancement in the field theory of electromagnetism which encompasses the description of dissipative media in addition to the well known field theory of electromagnetism in a non dissipative media.

Two versions have been formulated: The first one, which is not invariant under Lorentz transformation, has led to equations for both the electromagnetic fields (a slightly corrected set of Maxwell equations) , polarization and magnetization and, using a model for the sink variables, has unambiguously yielded energy densities, shown in equation (2.35) - equation (2.36) . These are positive and resemble the corresponding results in [14]. The results obtained are similar to the ones obtained in [1], however, they have been obtained here with a much simpler Lagrangian, containing fewer degrees of freedom. To be precise, two vector degrees of freedom (or six independent quantities) have been omitted from the beginning, thus making the formalism more useful for numerical calculations.

Secondly, a covariant Lagrangian was formulated, one that is invariant under the Lorentz transformation and contains spatial as well as temporal derivatives. From this, we have derived equations for both the electromagnetic fields, polarization and magnetization. Calculating the energy momentum tensor we have derived expression for both the energy density and the generalized Poynting vector. However, we find that with a specific model for the sink modes, the energy density derived from this covariant Lagrangian is not positive definite. This points to the need to introduce a better model for the sink degrees of freedom, perhaps with better physical underpinning.

Bibliography

[1] R. Englman, and A. Yahalom, "Energy density of a dissipative polarizable solid by a Lagrangean formalism," *Physics Letters A*, vol. 314, 367, 2003.

[2] A. Yahalom, R. Englman, and Y. Pinhasi, "Covariant formulation of the dynamics in a dissipative quantum dielectric obtained from a simplified Lagrangian," *Physics Letters A*, vol. 372, 2941-2948, 2008. Available: http://dx.doi.org/10.1016/j.physleta.2008.01.028 [Los-Alamos Archives - physics/0605060].

[3] G.A. Kluitenberg, and S.R. de Groot, "Systems with polarization and magnetization in an electromagnetic field," *Physica*, vol. 21, 146-168, 1955.

[4] R.M. Fano, L.J. Chu, and R.B. Adler, Electromagnetic Fields, Energy and Forces. MIT Press: Cambridge Mass., 1958. pp.453.

[5] B.U. Felderhof, "Relativistic hydrodynamics of magnetic and dielectric fluids in interaction with the electromagnetic field," *J. Chem. Phys.*, vol. 120, 3598, 2004.

[6] R. E. Rosensweig, "Continuum equations for magnetic and dielectric fluids with internal rotations," *J. Chem. Phys.*, vol. 121, 1228, 2004.

[7] G. W. Ford, M. Kac, and P. Mazur, "Statistical Mechanics of Assemblies of Coupled Oscillators," *J. Math. Phys.*, vol. 6, 504, 1965.

[8] S.R. de Groot, and P. Mazur, Irreversible Thermodynamics. North Holland: Amsterdam, 1958. pp.403.

[9] H. Bateman, "On Dissipative Systems and Related Variational Principles," *Phys Rev.*, vol. 38, 815, 1931.

[10] L.D. Landau, E.M. Lifshitz, and L.P. Pitaevskii, Electrodynamics of Continuous Media, 2nd edition. Pergamon Press: Oxford, 1984. Chapter IX.

[11] J.A. Stratton, Electromagnetic Theory. McGraw-Hill: New York, 1941. Chapter II.

[12] D.S. Jones, Theory of Electromagnetism. Pergamon Press: Oxford, 1964. section 2.20.

[13] J.D.Jackson, Classical Electrodynamics, Third Edition. Wiley: New York, 1999.

[14] R. Ruppin, "Electromagnetic energy density in a dispersive and absorptive material," *Phys. Lett. A*, vol. 299, 309, 2002.

[15] J.H. Hannay, "Comment on 'Hannay angle in an LCR circuit with time-dependent inductance, capacity and resistance'," *J. Phys. A: Math. Gen.*, vol. 35, 9699, 2002.

[16] J.D. Bekenstein, "Fine-structure constant variability, equivalence principle and cosmology," *Phys. Rev. D*, vol. 66, 123514, 2002.

[17] J.D. Bekenstein, "Fine-structure constant: Is it really a constant?," *Phys. Rev. D*, vol. 25, 1527, 1982.

[18] R. Loudon, "The propagation of electromagnetic energy through an absorbing dielectric," *J. Phys. A* , vol. 3, 233, 1970.

[19] J. Neufeld, "Reformulation of the Dielectric Constant and of Ohm's Law in an Absorbing Medium," *Phys. Rev.*, vol. 152, 708, 1966.

[20] J. Neufeld, "Wave propagation and group velocity in absorbing media," *Phys. Lett. A*, vol. 20, 69, 1969.

CHAPTER 3

Coupled-Mode Theory of Electromagnetism including Wide Band Distributed Interactions

Yosef Pinhasi[1], Yuri Lurie[2], Gad A. Pinhasi[3]

Abstract

Usually the electromagnetic field is naively thought as given by six independent quantities **E** and **B**. However, in chapter 2 the field was presented as dependent on a four-component vector potential. In this chapter a different approach is formulated, in which electromagnetic field is determined by two scalar complex functions. This approach will serve as a starting point for the following variational analysis given in the next chapter, where the electromagnetic field is shown to depend only on a single complex function.

[1]Ariel University Center of Samaria, P.O.B. 3, Ariel 40700 Israel, e-mail: yosip@ ariel.ac.il

[2]Ariel University Center of Samaria, P.O.B. 3, Ariel 40700 Israel, e-mail: ylurie@ ariel.ac.il

[3]Ariel University Center of Samaria, P.O.B. 3, Ariel 40700 Israel, e-mail: gadip@ ariel.ac.il

3.1 Introduction

Usually electromagnetic field is naively thought as given by six independent quantities **E** and **B**. However in chapter 2 the field was presented as dependent on a four-component vector potential. In this chapter a different approach is formulated, in which electromagnetic field is defined by two scalar complex functions. This approach will serve as a starting point for the following variational analysis given in chapter 4, where electromagnetic field is shown to depend on a single complex function only.

Space - frequency approach is used for the analysis of distributed interactions between electromagnetic fields and electric and magnetic currents sources. The theory is based on the presentation of the fields as a superposition of transverse eigenmodes and derivation of general scalar excitation equations for their amplitudes in the frequency domain. The approach is applied in the description of broadband phenomena occurring in distributed uniform cross-section medium, and characterized by a continuum of frequencies.

The approach taken here utilizes representation of the total electromagnetic fields and their sources in terms of vector functions, which are the eigenmode solutions of the medium, free of charge or current sources. First, a brief survey of the different type of eigenmode solutions is given, and important orthogonality relations are discussed. The orthogonality properties of the transverse modes enable the expansion of the transverse components of electric and magnetic fields excited in the waveguide, as a superposition of the eigenmodes. In the presence of electric and magnetic currents, the expansion yields scalar excitation equations for the modes of the medium in which the fields are excited

We start the discussion with the fundamental Maxwell's equations, transforming them to the frequency domain. The formalism is demonstrated by derivations for the electromagnetic field emitted by wiggling charged particles, while passing along an undulator of a waveguide-based free-electron laser. For this case, the radiation spectrum and the time-domain field are found as analytic expressions.

3.2 Maxwell's equations

The electromagnetic theory is governed by four partial differential equations derived by James Clerk Maxwell [1] in the 19th century. The Maxwell's equations describe the properties of the electric field $\mathbf{E}(\mathbf{r}, t)$ (in [V/m]) and the magnetic field $\mathbf{H}(\mathbf{r}, t)$ (in [A/m]) in space and time. These two vector fields are functions of the spatial coordinates $\mathbf{r}$ and the time t. In the followings, considering propagation of a charged beam through a uniform cross-section medium, we assume Cartesian (x, y, z) coordinates, where (x, y) are the transverse coordinates and z is the axis of propagation.

The fields are related to the sources, characterized by charge density $\rho(\mathbf{r}, t)$ (in [Coul/m^3]), current density $\mathbf{J}(\mathbf{r}, t)$ (in [A/m^2]), and magnetic current density $\mathbf{M}(\mathbf{r}, t)$ (in [V/m^2]) via Faraday's law of induction:

$$\nabla \times \mathbf{E} = -\frac{\partial \mathbf{B}}{\partial t} - \mathbf{M} \tag{3.1}$$

Ampere's law:

$$\nabla \times \mathbf{H} = \frac{\partial \mathbf{D}}{\partial t} + \mathbf{J} \tag{3.2}$$

and the two Gauss' laws:

$$\begin{aligned} \nabla \cdot \mathbf{D} &= \rho \\ \nabla \cdot \mathbf{B} &= 0 \end{aligned} \tag{3.3}$$

The electric displacement $\mathbf{D}(\mathbf{r}, t)$ is measured in [Coul/m^2] and the magnetic induction $\mathbf{B}(\mathbf{r}, t)$ is measured in [Wb/m^2] or in [Tesla]. Note that the current and charge densities relate each other via the continuity equation:

$$\nabla \cdot \mathbf{J} + \frac{\partial \rho}{\partial t} = 0 \tag{3.4}$$

The four Maxwell's equations together with the continuity equation form a set of dependent equations. Namely, one may refer to a one of the sources, e.g. to the current density $\mathbf{J}(\mathbf{r}, t)$ only, and express the charge density $\rho(\mathbf{r}, t)$ by using the continuity equation in terms of $\mathbf{J}(\mathbf{r}, t)$.

3.3 Presentation of the fields in the frequency domain

The electromagnetic field in the time domain is described by the space-time electric $\mathbf{E}(\mathbf{r}, t)$ and magnetic $\mathbf{H}(\mathbf{r}, t)$ signal vectors. It is convenient to present these fields in the frequency domain by using the Fourier transformation:

$$\mathbf{E}(\mathbf{r}, f) = \int_{-\infty}^{+\infty} \mathbf{E}(\mathbf{r}, t)\, e^{-j2\pi f t}\, \mathrm{d}t \tag{3.5}$$

where f denotes the frequency in [Hz]. Similar expression is defined for the Fourier transform $\mathbf{H}(\mathbf{r}, f)$ of the magnetic field. Since the electromagnetic signal is real (i.e. $\mathbf{E}^*(\mathbf{r}, t) = \mathbf{E}(\mathbf{r}, t)$), its Fourier transform satisfies $\mathbf{E}^*(\mathbf{r}, f) = \mathbf{E}(\mathbf{r}, -f)$. Analytic representation of the signal is given by the complex expression:

$$\widetilde{\mathbf{E}}(\mathbf{r}, t) \equiv \mathbf{E}(\mathbf{r}, t) + j\, \hat{\mathbf{E}}(\mathbf{r}, t) \tag{3.6}$$

where:

$$\hat{\mathbf{E}}(\mathbf{r}, t) = \frac{1}{\pi t} * \mathbf{E}(\mathbf{r}, t) = \frac{1}{\pi} \int_{-\infty}^{+\infty} \frac{\mathbf{E}(\mathbf{r}, t')}{t - t'}\, \mathrm{d}t' \tag{3.7}$$

is the Hilbert transform of $\mathbf{E}(\mathbf{r}, t)$. Fourier transformation of the analytic representation results in a 'phasor-like' function $\widetilde{\mathbf{E}}(\mathbf{r}, f)$ defined in the positive frequency domain and related to the Fourier transform by:

$$\widetilde{\mathbf{E}}(\mathbf{r}, f) = \begin{cases} 2\,\mathbf{E}(\mathbf{r}, f) & f > 0 \\ 0 & f < 0 \end{cases} \tag{3.8}$$

The Fourier transform can be decomposed in terms of the 'phasor like' functions according to:

$$\mathbf{E}(\mathbf{r}, f) = \frac{1}{2}\widetilde{\mathbf{E}}(\mathbf{r}, f) + \frac{1}{2}\widetilde{\mathbf{E}}^*(\mathbf{r}, -f) \tag{3.9}$$

and the inverse Fourier transform is then:

$$
\begin{aligned}
\mathbf{E}(\mathbf{r},t) &= \int_{-\infty}^{+\infty} \mathbf{E}(\mathbf{r},f)\, e^{+j2\pi ft}\, \mathrm{d}f = \\
&= \int_{-\infty}^{+\infty} \left[\frac{1}{2}\widetilde{\mathbf{E}}(\mathbf{r},f) + \frac{1}{2}\widetilde{\mathbf{E}}^{*}(\mathbf{r},-f)\right] e^{+j2\pi ft}\, \mathrm{d}f = \\
&= \frac{1}{2}\left\{\int_{0}^{\infty} \widetilde{\mathbf{E}}(\mathbf{r},f)\, e^{+j2\pi ft}\, \mathrm{d}f + \left[\int_{0}^{\infty} \widetilde{\mathbf{E}}(\mathbf{r},f)\, e^{+j2\pi ft}\, \mathrm{d}f\right]^{*}\right\} = \\
&= \Re\left\{\int_{0}^{\infty} \widetilde{\mathbf{E}}(\mathbf{r},f)\, e^{+j2\pi ft}\, \mathrm{d}f\right\}
\end{aligned}
\tag{3.10}
$$

Using the phasor-like notation for the fields and sources, Maxwell's equations can be rewritten in the frequency domain as:

$$
\begin{aligned}
&\nabla \times \widetilde{\mathbf{E}} = -j2\pi f\, \widetilde{\mathbf{B}} - \widetilde{\mathbf{M}} \\
&\nabla \times \widetilde{\mathbf{H}} = +j2\pi f\, \widetilde{\mathbf{D}} + \widetilde{\mathbf{J}} \\
&\nabla \cdot \widetilde{\mathbf{D}} = \widetilde{\rho} = -j2\pi f\, \nabla \cdot \widetilde{\mathbf{J}} \\
&\nabla \cdot \widetilde{\mathbf{B}} = 0
\end{aligned}
\tag{3.11}
$$

In the frequency domain, also the following relations exist:

$$
\begin{aligned}
\widetilde{\mathbf{D}}(\mathbf{r},f) &= \varepsilon(\mathbf{r},f)\, \widetilde{\mathbf{E}}(\mathbf{r},f) \\
\widetilde{\mathbf{B}}(\mathbf{r},f) &= \mu(\mathbf{r},f)\, \widetilde{\mathbf{H}}(\mathbf{r},f)
\end{aligned}
\tag{3.12}
$$

Here $\varepsilon(\mathbf{r},f)$ is the permittivity in [F/m] and $\mu(\mathbf{r},f)$ is the permeability in [H/m] of the medium in which the fields are propagating. In an isotropic, homogeneous, linear medium, where the permittivity $\varepsilon(f)$ and the permeability $\mu(f)$ are not spatially dependent, the Faraday's and Ampere's laws can be rewritten as:

$$
\begin{aligned}
\nabla \times \nabla \times \widetilde{\mathbf{E}} - k^2\widetilde{\mathbf{E}} &= -j2\pi f\mu \left[\widetilde{\mathbf{J}} + \frac{1}{j2\pi f\mu}\nabla \times \widetilde{\mathbf{M}}\right] \\
\nabla \times \nabla \times \widetilde{\mathbf{H}} - k^2\widetilde{\mathbf{H}} &= -j2\pi f\varepsilon \left[\widetilde{\mathbf{M}} + \frac{1}{-j2\pi f\varepsilon}\nabla \times \widetilde{\mathbf{J}}\right]
\end{aligned}
\tag{3.13}
$$

Where the propagation factor is:

$$k(f) \equiv 2\pi f \sqrt{\mu(f)\,\varepsilon(f)} \tag{3.14}$$

We identify $\widetilde{\mathbf{J}} + \nabla \times \widetilde{\mathbf{M}}/(j2\pi f \mu)$ as the total electric field source and $\widetilde{\mathbf{M}} + \nabla \times \widetilde{\mathbf{J}}/(-j2\pi f \varepsilon)$ as total the magnetic field source. Introducing the identity:

$$\nabla \times \nabla \times \begin{bmatrix} \widetilde{\mathbf{E}} \\ \widetilde{\mathbf{H}} \end{bmatrix} = \nabla\nabla \cdot \begin{bmatrix} \widetilde{\mathbf{E}} \\ \widetilde{\mathbf{H}} \end{bmatrix} - \nabla^2 \begin{bmatrix} \widetilde{\mathbf{E}} \\ \widetilde{\mathbf{H}} \end{bmatrix} \tag{3.15}$$

and Gauss' laws results in two partial differential equations:

$$\nabla^2 \widetilde{\mathbf{E}} + k^2 \widetilde{\mathbf{E}} = \underbrace{\frac{1}{\varepsilon}\nabla\widetilde{\rho}}_{-j\frac{2\pi f}{\varepsilon}\nabla\left(\nabla\cdot\widetilde{\mathbf{J}}\right)} + j2\pi f \mu \widetilde{\mathbf{J}} + \nabla \times \widetilde{\mathbf{M}} \tag{3.16}$$

$$\nabla^2 \widetilde{\mathbf{H}} + k^2 \widetilde{\mathbf{H}} = -\nabla \times \widetilde{\mathbf{J}} + j2\pi f \varepsilon \widetilde{\mathbf{M}}$$

These two independent expressions are the inhomogeneous Helmholtz equations, separately describing excitation of the electric and magnetic fields in the frequency domain.

3.4 The Wiener-Khinchine theorem and the Poynting's vector

The cross-correlation function of the time dependent electric $\mathbf{E}(\mathbf{r},t)$ and magnetic $\mathbf{H}(\mathbf{r},t)$ fields is given by:

$$R_{EH}(z,\tau) = \int_{-\infty}^{+\infty} \left\{ \iint \left[\mathbf{E}(\mathbf{r},t+\tau) \times \mathbf{H}(\mathbf{r},t)\right] \cdot \hat{\mathbf{z}}\, \mathrm{d}x\, \mathrm{d}y \right\} \mathrm{d}t \tag{3.17}$$

For finite energy signals, the total energy carried by the electromagnetic field is given by:

$$W(z) = R_{EH}(z,0) = \int_{-\infty}^{+\infty} \left\{ \iint \left[\mathbf{E}(\mathbf{r},t) \times \mathbf{H}(\mathbf{r},t)\right] \cdot \hat{\mathbf{z}}\, \mathrm{d}x\, \mathrm{d}y \right\} \mathrm{d}t \tag{3.18}$$

where we identify:

$$P(z,t) = \iint [\mathbf{E}(\mathbf{r},t) \times \mathbf{H}(\mathbf{r},t)] \cdot \hat{\mathbf{z}} \, \mathrm{d}x \, \mathrm{d}y \tag{3.19}$$

as the instantaneous power flow at a plane z. According to the Wiener-Khinchine theorem [2, 3], the spectral density of the electromagnetic energy is related to the cross-correlation function $R_{EH}(z,\tau)$ through the Fourier transformation:

$$\begin{aligned}
S_{EH}(z,f) &= \int_{-\infty}^{+\infty} R_{EH}(z,\tau) \; e^{-j2\pi f\tau} \, \mathrm{d}\tau = \\
&= \int_{-\infty}^{+\infty} \left\{ \int_{-\infty}^{+\infty} \left\{ \iint [\mathbf{E}(\mathbf{r},t+\tau) \times \mathbf{H}(\mathbf{r},t)] \cdot \hat{\mathbf{z}} \, \mathrm{d}x \, \mathrm{d}y \right\} \mathrm{d}t \right\} e^{-j2\pi f\tau} \, \mathrm{d}\tau = \\
&= \iint [\mathbf{E}(\mathbf{r},f) \times \mathbf{H}^*(\mathbf{r},f)] \cdot \hat{\mathbf{z}} \, \mathrm{d}x \, \mathrm{d}y
\end{aligned} \tag{3.20}$$

$S_{EH}(z,f)$ defines density spectrum of energy flux at plane z. It is a complex function satisfying $S_{EH}(z,-f) = S^*_{EH}(z,f)$. Consequently, the inverse Fourier transform can be expressed as:

$$\begin{aligned}
R_{EH}(z,\tau) &= \int_{-\infty}^{+\infty} S_{EH}(z,f) \; e^{+j2\pi f\tau} \, \mathrm{d}\tau = \\
&= \int_0^{\infty} S_{EH}(z,f) \; e^{+j2\pi f\tau} \, \mathrm{d}\tau \; + \left[\int_0^{\infty} S_{EH}(z,f) \; e^{+j2\pi f\tau} \, \mathrm{d}\tau \right]^* = \\
&= 2\Re \left\{ \int_0^{\infty} S_{EH}(z,f) \; e^{+j2\pi ft} \, \mathrm{d}f \right\}
\end{aligned} \tag{3.21}$$

Substituting in the expression of $S_{EH}(z,f)$ the Fourier transforms of the fields with their phasor notations:

$$\mathbf{E}(\mathbf{r},f)=\begin{cases}\frac{1}{2}\widetilde{\mathbf{E}}(\mathbf{r},f) & f>0\\ \frac{1}{2}\widetilde{\mathbf{E}}^*(\mathbf{r},-f) & f<0\end{cases}$$

$$\mathbf{H}^*(\mathbf{r},f)=\begin{cases}\frac{1}{2}\widetilde{\mathbf{H}}^*(\mathbf{r},f) & f>0\\ \frac{1}{2}\widetilde{\mathbf{H}}(\mathbf{r},-f) & f<0\end{cases} \tag{3.22}$$

gives:

$$S_{EH}(z,f)=\begin{cases}\frac{1}{4}\iint\left[\widetilde{\mathbf{E}}(\mathbf{r},f)\times\widetilde{\mathbf{H}}^*(\mathbf{r},f)\right]\cdot\hat{\mathbf{z}}\,\mathrm{d}x\,\mathrm{d}y & f>0\\ \frac{1}{4}\iint\left[\widetilde{\mathbf{E}}(\mathbf{r},-f)\times\widetilde{\mathbf{H}}^*(\mathbf{r},-f)\right]^*\cdot\hat{\mathbf{z}}\,\mathrm{d}x\,\mathrm{d}y & f<0\end{cases} \tag{3.23}$$

Following Parseval's theorem, the total energy carried by the electromagnetic field can also be calculated by integrating the spectral density $S_{EH}(z,f)$ over the entire frequency domain:

$$\begin{aligned}W(z)&=\int_{-\infty}^{+\infty}S_{EH}(z,f)\,\mathrm{d}f=\\&=\int_{0}^{+\infty}\frac{1}{4}\iint\left[\widetilde{\mathbf{E}}(\mathbf{r},f)\times\widetilde{\mathbf{H}}^*(\mathbf{r},f)\right]\cdot\hat{\mathbf{z}}\,\mathrm{d}x\,\mathrm{d}y\,\mathrm{d}f\\&\quad+\int_{-\infty}^{0}\frac{1}{4}\iint\left[\widetilde{\mathbf{E}}(\mathbf{r},-f)\times\widetilde{\mathbf{H}}^*(\mathbf{r},-f)\right]^*\cdot\hat{\mathbf{z}}\,\mathrm{d}x\,\mathrm{d}y\,\mathrm{d}f\\&=\int_{0}^{\infty}\left[\frac{1}{2}\iint\Re\left\{\widetilde{\mathbf{E}}(\mathbf{r},f)\times\widetilde{\mathbf{H}}^*(\mathbf{r},f)\right\}\cdot\hat{\mathbf{z}}\,\mathrm{d}x\,\mathrm{d}y\right]\mathrm{d}f\end{aligned} \tag{3.24}$$

We identify:

$$\begin{aligned}\frac{\mathrm{d}\,W(z)}{\mathrm{d}f}&=\frac{1}{2}\Re\left\{\iint\left[\widetilde{\mathbf{E}}(\mathbf{r},f)\times\widetilde{\mathbf{H}}^*(\mathbf{r},f)\right]\cdot\hat{\mathbf{z}}\,\mathrm{d}x\,\mathrm{d}y\right\}=\\&=\begin{cases}2\Re\{S_{EH}(z,f)\} & f>0\\ 0 & f<0\end{cases}\end{aligned} \tag{3.25}$$

as the spectral energy distribution (over positive frequencies) of the electromagnetic field (in [J/Hz] and:

$$\widetilde{\mathbf{S}}(\mathbf{r}, f) = \widetilde{\mathbf{E}}(\mathbf{r}, f) \times \widetilde{\mathbf{H}}^{*}(\mathbf{r}, f) \tag{3.26}$$

is the complex Poynting's vector [4]. Taking the divergent of the last expression and using Faraday's and Amperes laws results in:

$$\begin{aligned}
\nabla \cdot \widetilde{\mathbf{S}} &= \nabla \cdot \left(\widetilde{\mathbf{E}} \times \widetilde{\mathbf{H}}^{*}\right) = \widetilde{\mathbf{H}}^{*} \cdot \left(\nabla \times \widetilde{\mathbf{E}}\right) - \widetilde{\mathbf{E}} \cdot \left(\nabla \times \widetilde{\mathbf{H}}^{*}\right) = \\
&= \widetilde{\mathbf{H}}^{*} \cdot \left(-j2\pi f \widetilde{\mathbf{B}} - \widetilde{\mathbf{M}}\right) - \widetilde{\mathbf{E}} \cdot \left(+j2\pi f \widetilde{\mathbf{D}} + \widetilde{\mathbf{J}}\right)^{*} = \\
&= -j2\pi f \mu \left|\widetilde{\mathbf{H}}\right|^{2} - \widetilde{\mathbf{M}} \cdot \widetilde{\mathbf{H}}^{*} + j2\pi f \varepsilon^{*} \left|\widetilde{\mathbf{E}}\right|^{2} - \widetilde{\mathbf{J}}^{*} \cdot \widetilde{\mathbf{E}}
\end{aligned} \tag{3.27}$$

Employing the divergence theorem, the energy spectral density can be expressed in terms of the electric and magnetic current sources.

$$\begin{aligned}
&\frac{\mathrm{d}\, W(z)}{\mathrm{d}f} - \frac{\mathrm{d}\, W(z=0)}{\mathrm{d}f} = \\
&= \frac{1}{2}\Re\left\{\oiint \widetilde{\mathbf{S}}(\mathbf{r}, f) \cdot \hat{\mathbf{z}}\, \mathrm{d}x\, \mathrm{d}y\right\} = \\
&= \frac{1}{2}\Re\left\{\int_{0}^{z} \iint \nabla \cdot \widetilde{\mathbf{S}}(\mathbf{r}, f)\, \mathrm{d}x\, \mathrm{d}y\, \mathrm{d}z\right\} = \\
&= -\frac{1}{2}\Re\left\{\int_{0}^{z} \iint \left[\widetilde{\mathbf{J}}^{*} \cdot \widetilde{\mathbf{E}} + \widetilde{\mathbf{M}} \cdot \widetilde{\mathbf{H}}^{*}\right] \mathrm{d}x\, \mathrm{d}y\, \mathrm{d}z\right\}
\end{aligned} \tag{3.28}$$

3.5 Transverse eigen-modes

In a medium free of current and charge sources, the Helmholtz equations for the vector of electric $\widetilde{\mathbf{E}}(\mathbf{r}, f)$ and magnetic $\widetilde{\mathbf{H}}(\mathbf{r}, f)$ fields become homogeneous:

$$\begin{aligned}
\nabla^{2}\, \widetilde{\mathbf{E}} + k^{2}\, \widetilde{\mathbf{E}} &= 0 \\
\nabla^{2}\, \widetilde{\mathbf{H}} + k^{2}\, \widetilde{\mathbf{H}} &= 0
\end{aligned} \tag{3.29}$$

Since the Helmholtz equation is separable, the eigenmode solutions in an axially uniform medium correspond to waves propagating along the z axis:

$$\begin{aligned} \widetilde{\mathbf{E}}_{\pm q}(\mathbf{r}, f) &= \widetilde{\mathcal{E}}_{\pm q}(x, y)\; e^{\mp j\, k_{zq}\, z} \\ \widetilde{\mathbf{H}}_{\pm q}(\mathbf{r}, f) &= \widetilde{\mathcal{H}}_{\pm q}(x, y)\; e^{\mp j\, k_{zq}\, z} \end{aligned} \tag{3.30}$$

$\widetilde{\mathcal{E}}_{\pm q}(x, y)$ and $\widetilde{\mathcal{H}}_{\pm q}(x, y)$ are complex vectors representing the transverse profile and polarization of the electric and the magnetic fields of waveguide mode q, respectively. The sign plus stands for a forward mode going in the $+z$ direction, while the sign minus denotes a backward mode going in the $-z$ direction. The axial wavenumber of the mode is given by:

$$k_{zq} = \begin{cases} \sqrt{k^2 - k_{\perp q}^2} & k > k_{\perp q}\ , \quad \text{propagating modes} \\ -j\sqrt{k_{\perp q}^2 - k^2} & k < k_{\perp q}\ , \quad \text{cut} - \text{off modes} \end{cases} \tag{3.31}$$

The cut-off frequency of the q'th mode is $f_{co_q} = k_{\perp q} / \left(2\pi\sqrt{\mu\,\varepsilon}\right)$, thus:

$$k_{zq}(f) = \begin{cases} 2\pi\sqrt{\mu\,\varepsilon}\,\sqrt{f^2 - f_{co_q}^2} & f > f_{co_q} \quad \text{propagating modes} \\ -j\,2\pi\sqrt{\mu\,\varepsilon}\,\sqrt{f_{co_q}^2 - f^2} & f < f_{co_q} \quad \text{cut} - \text{off modes} \end{cases} \tag{3.32}$$

Above the cut-off ($f > f_{co_q}$), the axial wavenumber k_{zq} is a real quantity and the mode is propagating along the waveguide. The mode becomes evanescent for $f < f_{co_q}$.

3.5.1 Classification of mode solutions

For free-space and for transmission lines, the solution of interest is a transverse electromagnetic (TEM) wave with transverse components only. The electric field of the TEM modes can be found from the transverse gradient of a scalar function.

In waveguides and transmission lines with perfectly conducting walls, the boundary conditions can be satisfied also by two other types of modes; Transverse Electric (TE or H) modes, where all the components can be derived from the axial component of the magnetic field, and Transverse Magnetic (TM or E) modes, where all the components can be derived from the axial component of the electric field. A linear combinations of TE and TM modes always provide a complete and general solution.

Transverse Electro-Magnetic TEM modes

For these types of modes, the longitudinal components of both electric and magnetic fields are $\widetilde{\mathcal{E}}_{qz}(x,y) = \widetilde{\mathcal{H}}_{qz}(x,y) = 0$. Since the axial wavenumber of the TEM modes $k_{zq}(f) = k(f) = 2\pi f\sqrt{\mu\,\varepsilon}$, solutions of the transverse field components for each mode q can be found from:

$$\begin{aligned} \nabla_\perp^2 \widetilde{\mathcal{E}}_{q\perp} &= 0 \\ \nabla_\perp^2 \widetilde{\mathcal{H}}_{q\perp} &= 0 \end{aligned} \tag{3.33}$$

Alternatively, one can find a scalar potential function $\Phi(x,y)$, which is a solution of the two-dimensional Laplace equation:

$$\nabla_\perp^2 \Phi = 0 \tag{3.34}$$

which satisfies the proper boundary conditions. The electric field can then be expressed as the transverse gradient:

$$\widetilde{\mathcal{E}}_{q\perp} = -\nabla_\perp \Phi \tag{3.35}$$

and the magnetic field is given by:

$$\widetilde{\mathcal{H}}_{q\perp} = \frac{1}{\eta}\left(\hat{\mathbf{z}} \times \widetilde{\mathcal{E}}_{q\perp}\right) \tag{3.36}$$

where the $\eta(f) = \sqrt{\mu(f)/\varepsilon(f)}$ is the wave impedance in the medium.

Transverse Electric TE (or H) modes

For these modes the longitudinal component of the electric field $\widetilde{\mathcal{E}}_z(x,y) = 0$. The longitudinal component of the magnetic field is the solution of the homogeneous Helmholtz equation:

$$\nabla_\perp^2 \widetilde{\mathcal{H}}_{qz} + k_{\perp q}^2 \widetilde{\mathcal{H}}_{qz} = 0 \tag{3.37}$$

with the boundary condition: $\hat{\mathbf{n}} \cdot \nabla_\perp \widetilde{\mathcal{H}}_{qz} = 0$ at the waveguide walls. Using Maxwell's equations, the transverse component of the magnetic field is derived in terms of the longitudinal component:

$$\widetilde{\mathcal{H}}_{q\perp} = -j\frac{k_{zq}}{k_{\perp q}^2}\nabla_\perp \widetilde{\mathcal{H}}_{qz} \tag{3.38}$$

and the transverse component of the electric field is given by:

$$\widetilde{\mathcal{E}}_{q\perp} = -\eta_q\left(\hat{\mathbf{z}} \times \widetilde{\mathcal{H}}_{q\perp}\right) \tag{3.39}$$

where the TE mode wave impedance is $\eta_{TE_q}(f) = 2\pi f\mu/k_{zq}$.

Transverse Magnetic TM (or E) modes

For these modes the longitudinal component of the magnetic field $\widetilde{\mathcal{H}}_{qz}(x, y) = 0$. The longitudinal component of the electric field is the solution of the homogeneous Helmholtz equation:

$$\nabla_{\perp}^2 \widetilde{\mathcal{E}}_{qz} + k_{\perp q}^2 \widetilde{\mathcal{E}}_{qz} = 0 \tag{3.40}$$

with the boundary condition: $\widetilde{\mathcal{E}}_{qz} = 0$ at the waveguide walls.
Using Maxwell's equations, the transverse component of the electric field is derived in term of the longitudinal component:

$$\widetilde{\mathcal{E}}_{q\perp} = -j \frac{k_{zq}}{k_{\perp q}^2} \nabla_{\perp} \widetilde{\mathcal{E}}_{qz} \tag{3.41}$$

The transverse component of the magnetic field is given by:

$$\widetilde{\mathcal{H}}_{q\perp} = \frac{1}{\eta_q} \left(\hat{\mathbf{z}} \times \widetilde{\mathcal{E}}_{q\perp} \right) \tag{3.42}$$

where the TM mode wave impedance is $\eta_{TM_q}(f) = k_{zq}/2\pi f \varepsilon$.

3.5.2 Normalization of mode amplitudes and some orthogonalities

The field amplitudes are normalized via each mode's complex Poynting vector power:

$$\mathcal{N}_q = \iint \left[\widetilde{\mathcal{E}}_{q_\perp}(x, y) \times \widetilde{\mathcal{H}}_{q_\perp}^*(x, y) \right] \hat{\mathbf{z}} \, \mathrm{d}x \, \mathrm{d}y \tag{3.43}$$

The normalization $\mathcal{N}_q(f)$ is a frequency dependent quantity. It is real when $f > f_{co_q}$ (propagating modes) and imaginary for $f < f_{co_q}$ (evanescent modes). According to this normalization and by using the previously derived connections between the components, orthogonality relations between the transverse components of the eigenmodes can be written:

$$\begin{aligned} \iint \widetilde{\mathcal{E}}_{q_\perp}(x, y) \cdot \widetilde{\mathcal{E}}_{q'_\perp}^*(x, y) \, \mathrm{d}x \, \mathrm{d}y &= \eta_q^* \mathcal{N}_q \delta_{qq'} \\ \iint \widetilde{\mathcal{H}}_{q_\perp}(x, y) \cdot \widetilde{\mathcal{H}}_{q'_\perp}^*(x, y) \, \mathrm{d}x \, \mathrm{d}y &= \frac{1}{\eta_q} \mathcal{N}_q \delta_{qq'} \end{aligned} \tag{3.44}$$

where:

$$\delta_{qq'} = \begin{cases} 1 & q' = q \\ 0 & q' \neq q \end{cases} \tag{3.45}$$

The complex power normalization parameters $\mathcal{N}_q$ of the TE and TM modes can also be expressed in terms of the longitudinal component of the fields of the modes. This can be done by substituting the expressions for the transverse component of the fields into the orthogonality relation and imposing the boundary conditions at the waveguide walls:

$$\iint \widetilde{\mathcal{H}}_{qz}(x,y)\ \widetilde{\mathcal{H}}^*_{q'z}(x,y)\ \mathrm{d}x\ \mathrm{d}y = \frac{1}{\eta_{TEq}} \frac{k^2_{\perp q}}{|k_{zq}|^2} \mathcal{N}_q \delta_{qq'} \tag{3.46}$$

for the TE modes, and

$$\iint \widetilde{\mathcal{E}}_{qz}(x,y)\ \widetilde{\mathcal{E}}^*_{q'z}(x,y)\ \mathrm{d}x\ \mathrm{d}y = \eta^*_{TMq} \frac{k^2_{\perp q}}{|k_{zq}|^2} \mathcal{N}_q \delta_{qq'} \tag{3.47}$$

for the TM modes. As a result, the normalization can be expressed in terms of the axial components:

$$\mathcal{N}_q(f) = \begin{cases} \eta_{TEq} \frac{|k_{zq}|^2}{k^2_{\perp q}} \iint \left|\widetilde{\mathcal{H}}_{qz}(x,y)\right|^2 \mathrm{d}x\ \mathrm{d}y & \text{TE modes} \\ \frac{1}{\eta^*_{TMq}} \frac{|k_{zq}|^2}{k^2_{\perp q}} \iint \left|\widetilde{\mathcal{E}}_{qz}(x,y)\right|^2 \mathrm{d}x\ \mathrm{d}y & \text{TM modes} \end{cases} \tag{3.48}$$

Note that by symmetry considerations of Maxwell equations, it can be shown that the longitudinal components of the modes $\widetilde{\mathcal{E}}_{qz}(x,y)$ and $\widetilde{\mathcal{H}}_{qz}(x,y)$ can always be written as real functions (possibly multiplied by a constant complex number).

3.5.3 Relations between forward and backward modes

From Maxwell equations relations between forward and backward modes can be written:

$$\begin{aligned} \widetilde{\mathbf{E}}_{-q}(\mathbf{r},f) &= -\widetilde{\mathbf{E}}^*_{+q}(\mathbf{r},f) \\ \widetilde{\mathbf{H}}_{-q}(\mathbf{r},f) &= +\widetilde{\mathbf{H}}^*_{+q}(\mathbf{r},f) \end{aligned} \tag{3.49}$$

Given that the forward ($+q$) mode is written as:

$$\begin{aligned} \widetilde{\mathbf{E}}_{+q}(\mathbf{r},f) &= \left[\widetilde{\mathcal{E}}_{q\perp}(x,y) + \hat{\mathbf{z}}\ \widetilde{\mathcal{E}}_{qz}(x,y)\right]\ e^{-j\,k_{zq}\,z} \\ \widetilde{\mathbf{H}}_{+q}(\mathbf{r},f) &= \left[\widetilde{\mathcal{H}}_{q\perp}(x,y) + \hat{\mathbf{z}}\ \widetilde{\mathcal{H}}_{qz}(x,y)\right]\ e^{-j\,k_{zq}\,z} \end{aligned} \tag{3.50}$$

The backward $(-q)$ waveguide mode can be found from the above relations

$$\widetilde{\mathbf{E}}_{-q}(\mathbf{r},f) = \left[\widetilde{\mathcal{E}}_{q\perp}(x,y) - \hat{\mathbf{z}}\,\widetilde{\mathcal{E}}_{qz}(x,y)\right]\; e^{+j\,k_{zq}\,z}$$
$$\widetilde{\mathbf{H}}_{-q}(\mathbf{r},f) = \left[-\widetilde{\mathcal{H}}_{q\perp}(x,y) + \hat{\mathbf{z}}\,\widetilde{\mathcal{H}}_{qz}(x,y)\right]\; e^{+j\,k_{zq}\,z} \tag{3.51}$$

Important relations between the parameters of the forward and backward modes are given below.

Longitudinal wavenumber: $k_{z_{-q}}(f) = -k_{z_{+q}}(f)$

Mode impedance: $\eta_{-q}(f) = \eta_{+q}(f)$

Complex power normalization: $\mathcal{N}_{-q}(f) = -\mathcal{N}_{+q}(f)$.

3.6 Modal expansion of the fields

In this section we present the transformation procedure of the inhomogeneous, Helmholtz vector field equations into the scalar transmission line equations for an excited waveguide. The transverse component of the electric and magnetic fields in a uniform cross-section medium can be expanded in any cross-section as a linear superposition of a complete set of transverse eigenmodes [5]:

$$\widetilde{\mathbf{E}}_{\perp}(\mathbf{r},f) = \sum_q \widetilde{V}_q(z,f)\,\widetilde{\mathcal{E}}_{q\perp}(x,y)$$
$$\widetilde{\mathbf{H}}_{\perp}(\mathbf{r},f) = \sum_q \widetilde{I}_q(z,f)\,\widetilde{\mathcal{H}}_{q\perp}(x,y) \tag{3.52}$$

The summations include propagating and cut-off TE and TM modes, for which $\widetilde{V}_q(z,f)$ and $\widetilde{I}_q(z,f)$ are the scalar amplitude of the electric and magnetic fields respectively in positive frequencies, with field profiles $\widetilde{\mathcal{E}}_{q\perp}(x,y)$ and $\widetilde{\mathcal{H}}_{q\perp}(x,y)$ respectively. Inner product with the modes profiles and employing the orthogonality relations results in:

$$\widetilde{V}_q(z,f) = \frac{1}{\eta_q^* \mathcal{N}_q} \iint \widetilde{\mathbf{E}}_{\perp}(\mathbf{r},f)\cdot\widetilde{\mathcal{E}}_{q_\perp}^*(x,y)\,\mathrm{d}x\,\mathrm{d}y$$
$$\widetilde{I}_q(z,f) = \frac{\eta_q}{\mathcal{N}_q} \iint \widetilde{\mathbf{H}}_{\perp}(\mathbf{r},f)\cdot\widetilde{\mathcal{H}}_{q_\perp}^*(x,y)\,\mathrm{d}x\,\mathrm{d}y \tag{3.53}$$

These scalar, frequency dependent quantities can be regarded as voltage and current phasors. Expressions for the longitudinal component of the electric and magnetic fields are obtained after substituting the modal representation of the transverse fields into Maxwell's equations, resulting in:

$$\begin{aligned}\widetilde{E}_z(\mathbf{r},f) &= \sum_q \widetilde{I}_q(z,f)\,\widetilde{\mathcal{E}}_{qz}(x,y) - \frac{1}{j2\pi f\,\varepsilon}\widetilde{J}_z(\mathbf{r},f)\\ \widetilde{H}_z(\mathbf{r},f) &= \sum_q \widetilde{V}_q(z,f)\,\widetilde{\mathcal{H}}_{qz}(x,y) - \frac{1}{j2\pi f\,\mu}\widetilde{M}_z(\mathbf{r},f)\end{aligned} \tag{3.54}$$

3.6.1 The telegrapher equations for the excited waveguide

By imposing the appropriate boundary conditions, the Maxwell vector equations are transformed into scalar coupled differential equations, which describe the evolution of the equivalent voltage $\widetilde{V}_q(z,f)$ and current $\widetilde{I}_q(z,f)$ amplitudes:

$$\begin{aligned}&\frac{\mathrm{d}}{\mathrm{d}z}\widetilde{V}_q(z,f) + jk_{zq}\widetilde{I}_q(z,f) =\\ &= -\frac{1}{\mathcal{N}_q}\left[\iint \widetilde{J}_z(\mathbf{r},f)\,\mathcal{E}^*_{qz}(x,y)\,\mathrm{d}x\,\mathrm{d}y \;+\; \iint \widetilde{\mathbf{M}}_\perp(\mathbf{r},f)\cdot\widetilde{\mathcal{H}}^*_{q\perp}(x,y)\,\mathrm{d}x\,\mathrm{d}y\right]\\[1em] &\frac{\mathrm{d}}{\mathrm{d}z}\widetilde{I}_q(z,f) + jk_{zq}\widetilde{V}_q(z,f) =\\ &= -\frac{1}{\mathcal{N}_q}\frac{\eta_q}{\eta_q^*}\left[\iint \widetilde{\mathbf{J}}_\perp(\mathbf{r},f)\cdot\widetilde{\mathcal{E}}^*_{q\perp}(x,y)\,\mathrm{d}x\,\mathrm{d}y \;+\; \iint \widetilde{M}_z(\mathbf{r},f)\,\widetilde{\mathcal{H}}^*_{qz}(x,y)\,\mathrm{d}x\,\mathrm{d}y\right]\end{aligned} \tag{3.55}$$

These are known as the telegrapher transmission-line equations, which can also be written in a separated form:

$$
\begin{aligned}
&\frac{\mathrm{d}^2}{\mathrm{d}z^2}\widetilde{V}_q(z,f) + k_{zq}^2\widetilde{V}_q(z,f) = \\
&= -\frac{\mathrm{d}}{\mathrm{d}z}\left\{\frac{1}{\mathcal{N}_q}\left[\iint \widetilde{J}_z(\mathbf{r},f)\,\mathcal{E}_{qz}^*(x,y)\,\mathrm{d}x\mathrm{d}y\right.\right.\\
&\left.\left. + \iint \widetilde{\mathbf{M}}_\perp(\mathbf{r},f)\cdot\widetilde{\mathcal{H}}_{q_\perp}^*(x,y)\,\mathrm{d}x\mathrm{d}y\right]\right\}\\
&+\ jk_{zq}\frac{1}{\mathcal{N}_q}\frac{\eta_q}{\eta_q^*}\left[\iint \widetilde{\mathbf{J}}_\perp(\mathbf{r},f)\cdot\widetilde{\mathcal{E}}_{q_\perp}^*(x,y)\,\mathrm{d}x\mathrm{d}y + \iint \widetilde{M}_z(\mathbf{r},f)\,\widetilde{\mathcal{H}}_{qz}^*(x,y)\,\mathrm{d}x\mathrm{d}y\right]
\end{aligned}
$$

$$
\begin{aligned}
&\frac{\mathrm{d}^2}{\mathrm{d}z^2}\widetilde{I}_q(z,f) + k_{zq}^2\widetilde{I}_q(z,f) = \\
&= jk_{zq}\frac{1}{\mathcal{N}_q}\left[\iint \widetilde{J}_z(\mathbf{r},f)\,\mathcal{E}_{qz}^*(x,y)\,\mathrm{d}x\mathrm{d}y + \iint \widetilde{\mathbf{M}}_\perp(\mathbf{r},f)\cdot\widetilde{\mathcal{H}}_{q_\perp}^*(x,y)\,\mathrm{d}x\mathrm{d}y\right]\\
&-\ \frac{\mathrm{d}}{\mathrm{d}z}\left\{\frac{1}{\mathcal{N}_q}\frac{\eta_q}{\eta_q^*}\left[\iint \widetilde{\mathbf{J}}_\perp(\mathbf{r},f)\cdot\widetilde{\mathcal{E}}_{q_\perp}^*(x,y)\mathrm{d}x\mathrm{d}y\right.\right.\\
&\left.\left. + \iint \widetilde{M}_z(\mathbf{r},f)\,\widetilde{\mathcal{H}}_{qz}^*(x,y)\mathrm{d}x\mathrm{d}y\right]\right\}
\end{aligned}
\tag{3.56}
$$

3.6.2 Energy spectrum

The energy spectrum can now be written in terms of the mode amplitudes:

$$
\begin{aligned}
\frac{\mathrm{d}\,W(z)}{\mathrm{d}f} &= \frac{1}{2}\Re\left\{\iint\left[\widetilde{\mathbf{E}}(\mathbf{r},f)\times\widetilde{H}^*(\mathbf{r},f)\right]\cdot\hat{\mathbf{z}}\,\mathrm{d}x\,\mathrm{d}y\right\} = \\
&= \frac{1}{2}\Re\left\{\iint\left(\left[\sum_q \widetilde{V}_q(z,f)\,\widetilde{\mathcal{E}}_{q\perp}(x,y)\right]\right.\right.\\
&\qquad\left.\left.\times\left[\sum_q \widetilde{I}_q^*(z,f)\,\widetilde{\mathcal{H}}_{q\perp}^*(x,y)\right]\right)\cdot\hat{\mathbf{z}}\,\mathrm{d}x\,\mathrm{d}y\right\} = \\
&= \frac{1}{2}\Re\left\{\sum_q\sum_{q'}\widetilde{V}_q(z,f)\,\widetilde{I}_{q'}^*(z,f)\iint\left[\widetilde{\mathcal{E}}_{q\perp}(x,y)\times\widetilde{\mathcal{H}}_{q'\perp}^*(x,y)\right]\cdot\hat{\mathbf{z}}\,\mathrm{d}x\,\mathrm{d}y\right\}
\end{aligned}
\tag{3.57}
$$

Using the orthogonality relation between the modes leads to:

$$\frac{\mathrm{d}\,W(z)}{\mathrm{d}f} = \frac{1}{2}\Re\left\{\sum_q \widetilde{V}_q(z,f)\,\widetilde{I}_q^*(z,f)\,\mathcal{N}_q(f)\right\} \tag{3.58}$$

Since the normalization $\mathcal{N}_q(f)$ is real for propagating modes and imaginary for evanescent ones, we can write the last expression in the form:

$$\begin{aligned}\frac{\mathrm{d}\,W(z)}{\mathrm{d}f} &= \frac{1}{2}\sum_{\substack{q\\ \text{Propagating}\\ \text{modes}}} \Re\left\{\widetilde{V}_q(z,f)\,\widetilde{I}_q^*(z,f)\right\}\,\Re\left\{\mathcal{N}_q(f)\right\}\\ &\quad -\frac{1}{2}\sum_{\substack{q\\ \text{Cut-off}\\ \text{modes}}} \Im\left\{\widetilde{V}_q(z,f)\,\widetilde{I}_q^*(z,f)\right\}\,\Im\left\{\mathcal{N}_q(f)\right\}\end{aligned} \tag{3.59}$$

3.7 Incident and reflected waves

The scalar, frequency-dependent amplitudes $\widetilde{V}_q(z,f)$ and $\widetilde{I}_q(z,f)$ are fast varying quantities along the axial dimension z. It is possible to arrange the telegrapher equations to describe excitation of forward and backward waves by defining:

$$\begin{aligned} C_{+q}(z,f) &= \frac{1}{2}\left[\widetilde{V}_q(z,f) + \widetilde{I}_q(z,f)\right]\; e^{-j\,k_{zq}\,z}\\ C_{-q}(z,f) &= \frac{1}{2}\left[\widetilde{V}_q(z,f) - \widetilde{I}_q(z,f)\right]\; e^{+j\,k_{zq}\,z}\end{aligned} \tag{3.60}$$

In this case one derives:

$$\begin{aligned} \widetilde{V}_q(z,f) &= C_{+q}(z,f)\; e^{-j\,k_{zq}\,z} + C_{-q}(z,f)\; e^{+j\,k_{zq}\,z}\\ \widetilde{I}_q(z,f) &= C_{+q}(z,f)\; e^{-j\,k_{zq}\,z} - C_{-q}(z,f)\; e^{+j\,k_{zq}\,z}\end{aligned} \tag{3.61}$$

The perpendicular components of the electric and magnetic field can now be written as a linear superposition of a complete set of transverse eigenmodes, in terms of the amplitudes $C_{+q}(z,f)$ and $C_{-q}(z,f)$:

$$\widetilde{E}_{\perp}(\mathbf{r},f) = \sum_q \left[C_{+q}(z,f)\, e^{-j\,k_{zq}\,z} + C_{-q}(z,f)\, e^{+j\,k_{zq}\,z}\right]\, \widetilde{\mathcal{E}}_{q\perp}(x,y) =$$
$$= \sum_q \left[C_{+q}(z,f)\, \widetilde{\mathcal{E}}_{+q\perp}(x,y)\, e^{-j\,k_{zq}\,z} + C_{-q}(z,f)\, \widetilde{\mathcal{E}}_{-q\perp}(x,y)\, e^{+j\,k_{zq}\,z}\right]$$

$$\widetilde{H}_{\perp}(\mathbf{r},f) = \sum_q \left[C_{+q}(z,f)\, e^{-j\,k_{zq}\,z} - C_{-q}(z,f)\, e^{+j\,k_{zq}\,z}\right]\, \widetilde{\mathcal{H}}_{q\perp}(x,y) =$$
$$= \sum_q \left[C_{+q}(z,f)\, \widetilde{\mathcal{H}}_{+q\perp}(x,y)\, e^{-j\,k_{zq}\,z} + C_{-q}(z,f)\, \widetilde{\mathcal{H}}_{-q\perp}(x,y)\, e^{+j\,k_{zq}\,z}\right] \tag{3.62}$$

Similarly, expressions for the longitudinal component of the electric and magnetic fields can be re-written as:

$$\widetilde{E}_z(\mathbf{r},f) =$$
$$= \sum_q \left[C_{+q}(z,f)\, e^{-j\,k_{zq}\,z} - C_{-q}(z,f)\, e^{+j\,k_{zq}\,z}\right]\, \widetilde{\mathcal{E}}_{qz}(x,y) - \frac{1}{j2\pi\, f\, \varepsilon}\, \widetilde{J}_z(\mathbf{r},f)$$
$$= \sum_q \left[C_{+q}(z,f)\, \widetilde{\mathcal{E}}_{+qz}(x,y)\, e^{-j\,k_{zq}\,z} + C_{-q}(z,f)\, \widetilde{\mathcal{E}}_{-qz}(x,y)\, e^{+j\,k_{zq}\,z}\right]$$
$$- \frac{j}{j2\pi\, f\, \varepsilon} \widetilde{J}_z(\mathbf{r},f)$$

$$\breve{\widetilde{H}}_z(\mathbf{r},f) =$$
$$= \sum_q \left[C_{+q}(z,f)\, e^{-j\,k_{zq}\,z} + C_{-q}(z,f)\, e^{+j\,k_{zq}\,z}\right]\, \widetilde{\mathcal{H}}_{qz}(x,y) - \frac{j}{2\pi\, f\, \mu} \widetilde{M}_z(\mathbf{r},f)$$
$$= \sum_q \left[C_{+q}(z,f)\, \widetilde{\mathcal{H}}_{+qz}(x,y)\, e^{-j\,k_{zq}\,z} + C_{-q}(z,f)\, \widetilde{\mathcal{H}}_{-qz}(x,y)\, e^{+j\,k_{zq}\,z}\right]$$
$$- \frac{j}{2\pi\, f\, \mu} \widetilde{M}_z(\mathbf{r},f) \tag{3.63}$$

3.7.1 Excitation equations

The evolution of the amplitudes of the excited modes is described by a set of coupled differential equations for forward modes:

$$\begin{aligned}
&\frac{\mathrm{d}}{\mathrm{d}z} C_{+q}(z,f) = \\
&= -\frac{1}{2\mathcal{N}_q} e^{+j\,k_{zq}\,z} \iint \left[\frac{\eta_q}{\eta_q^*}\widetilde{\mathbf{J}}_\perp(\mathbf{r},f) + \hat{\mathbf{z}}\,\widetilde{J}_z(\mathbf{r},f)\right] \cdot \widetilde{\mathcal{E}}^*_{+q}(x,y)\,\mathrm{d}x\,\mathrm{d}y \\
&\quad - \frac{1}{2\mathcal{N}_q} e^{+j\,k_{zq}\,z} \iint \left[\widetilde{\mathbf{M}}_\perp(\mathbf{r},f) + \hat{\mathbf{z}}\,\frac{\eta_q}{\eta_q^*}\widetilde{M}_z(\mathbf{r},f)\right] \cdot \widetilde{\mathcal{H}}^*_{+q}(x,y)\,\mathrm{d}x\,\mathrm{d}y
\end{aligned} \tag{3.64}$$

and for backward modes:

$$\begin{aligned}
&\frac{\mathrm{d}}{\mathrm{d}z} C_{-q}(z,f) = \\
&= +\frac{1}{2\mathcal{N}_q} e^{-j\,k_{zq}\,z} \iint \left[\frac{\eta_q}{\eta_q^*}\widetilde{\mathbf{J}}_\perp(\mathbf{r},f) + \hat{\mathbf{z}}\,\widetilde{J}_z(\mathbf{r},f)\right] \cdot \widetilde{\mathcal{E}}^*_{-q}(x,y)\,\mathrm{d}x\,\mathrm{d}y \\
&\quad + \frac{1}{2\mathcal{N}_q} e^{-j\,k_{zq}\,z} \iint \left[\widetilde{M}_\perp(\mathbf{r},f) + \hat{\mathbf{z}}\,\frac{\eta_q}{\eta_q^*}\,\widetilde{M}_z(\mathbf{r},f)\right] \cdot \widetilde{\mathcal{H}}^*_{-q}(x,y)\,\mathrm{d}x\,\mathrm{d}y
\end{aligned} \tag{3.65}$$

Note the ratio:

$$\frac{\eta_q}{\eta_q^*} = \begin{cases} +1 & \text{for propagating modes} \\ -1 & \text{for cut} - \text{off modes} \end{cases} \tag{3.66}$$

3.7.2 Energy flow

Substituting the modal expansion into the expression for the spectral energy distribution results in a summation of energy spectrums of the excited modes:

$$
\begin{aligned}
\frac{\mathrm{d}\,W(z)}{\mathrm{d}f} &= \frac{1}{2}\Re\left\{\sum_q \widetilde{V}_q(z,f)\ \widetilde{I}^*_q(z,f)\ \mathcal{N}_q(f)\right\} = \\
&= \frac{1}{2}\,\Re\left\{\sum_q \left[C_{+q}(z,f)\ e^{-j\,k_{zq}\,z} + C_{-q}(z,f)\ e^{+j\,k_{zq}\,z}\right]\right. \\
&\qquad \left.\cdot\left[C_{+q}(z,f)\ e^{-j\,k_{zq}\,z} - C_{-q}(z,f)\ e^{+j\,k_{zq}\,z}\right]^*\ \mathcal{N}_q(f)\right\} = \\
&= \frac{1}{2}\,\Re\left\{\sum_q \left[|C_{+q}(z,f)|^2\ e^{-j\left(k_{zq}-k^*_{zq}\right)z} - |C_{-q}(z,f)|^2\ e^{+j\left(k_{zq}-k^*_{zq}\right)z}\right.\right. \\
&\qquad -C_{+q}(z,f)\ C^*_{-q}(z,f)\ e^{-j\left(k_{zq}+k^*_{zq}\right)z} \\
&\qquad \left.\left.+C^*_{+q}(z,f)\ C_{-q}(z,f)\ e^{+j\left(k_{zq}+k^*_{zq}\right)z}\right]\ \mathcal{N}_q(f)\right\} = \\
&= \frac{1}{2}\,\Re\left\{\sum_q \left[|C_{+q}(z,f)|^2\ e^{-2\Im\{k_{zq}\}\,z} - |C_{-q}(z,f)|^2\ e^{+2\Im\{k_{zq}\}\,z}\right.\right. \\
&\qquad -C_{+q}(z,f)\ C^*_{-q}(z,f)\ e^{-j\,2\Re\{k_{zq}\}\,z} \\
&\qquad \left.\left.+C^*_{+q}(z,f)\ C_{-q}(z,f)\ e^{-j\,2\Re\{k_{zq}\}\,z}\right]\ \mathcal{N}_q(f)\right\} = \\
&= \frac{1}{2}\,\Re\left\{\sum_q \left[|C_{+q}(z,f)|^2\ e^{-2\Im\{k_{zq}\}\,z} - |C_{-q}(z,f)|^2\ e^{+2\Im\{k_{zq}\}\,z}\right.\right. \\
&\qquad \left.\left.+\ j\,2\Im\left\{C_{+q}(z,f)\ C^*_{-q}(z,f)\ e^{-j\,2\Re\{k_{zq}\}\,z}\right\}\right]\ \mathcal{N}_q(f)\right\}
\end{aligned}
\tag{3.67}
$$

Remembering that k_{zq} is real for propagating modes and imaginary for cut-off modes, the last derivation can be written as:

$$\frac{\mathrm{d}\,W(z)}{\mathrm{d}f} = \frac{1}{2} \sum_{\substack{q \\ \text{Propagating} \\ \text{modes}}} \left[|C_{+q}(z,f)|^2 - |C_{-q}(z,f)|^2 \right] \Re\{\mathcal{N}_q\} + \sum_{\substack{q \\ \text{Cut-off} \\ \text{modes}}} \Im\left\{ C_{+q}(z,f)\, C^*_{-q}(z,f) \right\} \Im\{\mathcal{N}_q\} \tag{3.68}$$

3.8 Electromagnetic radiation emitted by wiggling charged particles

The described approach is applied to description of electromagnetic radiation, emitted by wiggling accelerated charged particles in a waveguide structure. Such emission takes place in free-electron lasers; some results of analytic and numeric analysis of free-electron lasers in different linear and non-linear regimes can be found in [6]-[12].

Electrons passing through a magnetic undulator emit a partially coherent radiation which is called *undulator synchrotron radiation.* In a classical analysis each electron is a point source, which can be treated as a moving radiating dipole while wiggling. Since the radiation process takes place in the absence of applied electromagnetic radiation, it can be termed *spontaneous emission.*

The radiation of each individual electron adds up coherently as it advances and radiates along the undulator. However, when a continuous (unmodulated and unbunched) electron beam advances through a periodic field of a wiggler, the radiation from different electrons, which enter the undulator at random, adds up incoherently. This radiation is essentially *shot noise*, resulting from the charge fluctuations in the e-beam due to its corpuscular nature and the random electron distribution.

Coherent summation of the radiation from individual particles occurs when electrons are inserted in a single short bunch (shorter than the oscillation period of the emitted radiation) or enter as a periodic train of bunches at the frequency of the emitted radiation. Only in these cases do the electrons radiate in phase with each other (*super-radiant emission*) and the radiation is coherent [8].

3.8.1 Excitation of the electromagnetic radiation by a single electron

A single electron moving at an instantaneous velocity $\mathbf{v}_i$ is a point source with current density $\mathbf{J}_i(\mathbf{r}, t)$ given in the space-time domain by:

$$\mathbf{J}_i(\mathbf{r}, t) = -e\mathbf{v}_i \delta(x - x_i)\delta(y - y_i)\delta[z - z_i(t)] \tag{3.69}$$

where $-e$ is the electric charge of the electron and (x_i, y_i, z_i) are the coordinates of the electron's position at a time t. It is convenient to take z, the coordinate of axial propagation, as the independent variable, and to write the time it takes a particle to arrive at a point z as:

$$t_i(z) = t_{0i} + \int_0^z \frac{\mathrm{d}z'}{v_{zi}(z')} \tag{3.70}$$

t_{0i} is the time that the particle entered at $z = 0$ and $v_{zi}(z)$ is its axial velocity along the path of motion. To find the electromagnetic field radiated from the moving charge, it is first required to calculate the Fourier transform of the current density as given in (3.69):

$$\mathbf{J}_i(\mathbf{r}, f) \equiv \int_{-\infty}^{+\infty} \mathbf{J}(\mathbf{r}, t) e^{-j2\pi f t}\, \mathrm{d}t = -e\, \frac{\mathbf{v}_i}{v_{zi}}\, \delta(x - x_i)\, \delta(y - y_i)\, e^{-j2\pi f t_i(z)} \tag{3.71}$$

and substitute the 'phasor like' component $\widetilde{\mathbf{J}}_i(\mathbf{r}, f) = 2\mathbf{J}_i(\mathbf{r}, f)u(f)$ where $u(f)$ is the step function, into the excitation equations (3.64), (3.65):

$$\frac{\mathrm{d}}{\mathrm{d}z}\widetilde{C}_{\pm q}(z, f) = \pm \frac{e}{\mathcal{N}_q} \frac{1}{v_{zi}} \mathbf{v}_i \cdot \widetilde{\mathcal{E}}^*_{\pm q}(x_i, y_i) e^{-j[2\pi f t_i(z) \mp k_{zq} z]} \tag{3.72}$$

(only the propagating waves are considered here).

When an electron passes through the periodic field of an undulator, its total velocity vector $\mathbf{v}_i$ consists of a transverse wiggling component of amplitude $\widetilde{\mathcal{V}}^w_\perp$, which is due to the Lorentz force, in addition to a longitudinal axial velocity v_{zi}:

$$\mathbf{v}_i(z) = \hat{\mathbf{z}} v_{zi}(z) + \Re\left\{\widetilde{\mathcal{V}}^w_\perp e^{+jk_w z}\right\} \tag{3.73}$$

$k_w = 2\pi/\lambda_w$ where λ_w is the wiggler period. The electron transverse trajectory in the wiggler is given by:

$$\mathbf{r}_{\perp i}(z) = \bar{\mathbf{r}}_{\perp i} + \Re\left\{\widetilde{\mathbf{r}}_{\perp w} e^{+jk_w z}\right\} \tag{3.74}$$

where $\bar{\mathbf{r}}_{\perp i} = (\bar{x}_i, \bar{y}_i)$ describes the average (over a wiggling period) transverse coordinates of the electron and $\widetilde{\mathbf{r}}_{\perp w} = j\frac{\widetilde{\mathcal{V}}_\perp^w}{k_w v_{z0}}$ is the amplitude of transverse displacement of the wiggling electron trajectory ('quiver').

Because of the finite extent of the 'quiver' amplitude of the wiggling electron, the variation of the electric field in the transverse dimension has to be taken into account. To find the field distribution at a plane z along the path of the electrons, we expand the local electric field profile around the axis of the average electron trajectory $(\bar{x}_i, \bar{y}_i)$ in a Taylor series:

$$\widetilde{\mathcal{E}}_q[(x_i(z), y_i(z)] \simeq \widetilde{\mathcal{E}}_q(\bar{x}_i, \bar{y}_i) + \left[\Re\left\{\widetilde{\mathbf{r}}_{\perp w} e^{+jk_w z}\right\} \cdot \nabla_\perp\right] \widetilde{\mathcal{E}}_q(x, y)\Big|_{\bar{x}_i, \bar{y}_i} \tag{3.75}$$

The amplitude of the propagating mode q of radiation in the frequency domain is found after substitution of the phase-matched components of the velocity (3.73) and field (3.75) product, into Eq. (3.72):

$$\begin{aligned}&\frac{\mathrm{d}}{\mathrm{d}z}\widetilde{C}_{\pm q}(z, f) = \\ &= \pm\frac{e}{2\mathcal{N}_q}\frac{1}{v_{zi}} e^{-j[2\pi f t_i(z) \mp (k_{zq}+k_w)z]} \left[\widetilde{\mathcal{V}}_\perp^w + \hat{\mathbf{z}} v_{z0}(\widetilde{\mathbf{r}}_{w\perp}\cdot\nabla_\perp)\right] \cdot \widetilde{\mathcal{E}}_{\pm q}^*(x, y)\Big|_{\bar{x}_i, \bar{y}_i}\end{aligned} \tag{3.76}$$

Remembering that for TM modes $\nabla_\perp \widetilde{\mathcal{E}}_{qz} = j\frac{k_{\perp q}^2}{k_{zq}}\widetilde{\mathcal{E}}_{q\perp}$ and using the expression (3.70) for $t_i(z)$, equation (3.76) can be written:

$$\frac{\mathrm{d}}{\mathrm{d}z}\widetilde{C}_{\pm q}(z, f) = \pm\frac{e\zeta_q}{2\mathcal{N}_q}\frac{1}{v_{zi}}\, \widetilde{\mathcal{V}}_\perp^w \cdot \widetilde{\mathcal{E}}_{\pm q\perp}^*(\bar{x}_i, \bar{y}_i)\, e^{-j\left[2\pi f t_{0i} \mp \int_0^z \theta_q(z')dz'\right]} \tag{3.77}$$

where we define

$$\zeta_q \equiv \begin{cases} 1 & \text{for } TE \text{ modes} \\ 1 - \frac{k_{\perp q}^2}{k_{zq}k_w} & \text{for } TM \text{ modes}\end{cases} \tag{3.78}$$

and

$$\theta_q(z) \equiv \frac{2\pi f}{v_{zi}(z)} - (k_{zq} + k_w) \tag{3.79}$$

is the detuning parameter at point z.

3.8.2 Spontaneous emission in free-electron lasers employing waveguides

When an electron beam passes through an undulator, each electron emits an undulator synchrotron radiation. The total field is given by the summation of all the contributions of individual electrons to the radiation. Neglecting the effect of the radiation on the electrons, it is assumed that the electrons in the beam all move at a constant (averaged over wiggler period) axial velocity $v_{zi}(z) = v_{z0}$ keeping their initial detuning parameter θ_q constant along the wiggler. Consequently, the solution of the excitation equation (3.77) at the exit of a wiggler of length L_w is found to be:

$$\widetilde{C}_{+q}(L_w, f) = \mathcal{A}_q^{sp} \operatorname{sinc}\left(\frac{1}{2}\theta_q L_w\right) e^{-j\left(2\pi f t_{0i}+\frac{1}{2}\theta_q L_w\right)} \tag{3.80}$$

where $\mathcal{A}_q^{sp} \equiv \frac{e\zeta_q}{2N_q}\frac{L_w}{v_{z0}}\widetilde{\mathcal{V}}_\perp^w \cdot \widetilde{\mathcal{E}}_{+q\perp}^*(\bar{x}_i, \bar{y}_i)$ and $\operatorname{sinc}(x) \equiv \sin(x)/x$. The backward wave is supposed to be negligible at the plain $z = L_w$, i.e. $\widetilde{C}_{-q}(L_w, f) = 0$, what corresponds no external electromagnetic wave entering the waveguide through this plain. Inspection of this equation reveals that the field amplitude $\widetilde{C}_{+q}(L_w, f)$ has a significant value only within a frequency range where a condition of phase matching $\theta_q \approx 0$ is obtained. This corresponds to the synchronism condition $v_p \simeq v_{z0}$, where $v_p = 2\pi f_s/(k_{zq} + k_w)$ is the phase velocity of the ponderomotive wave. Explicit solutions for synchronism frequencies f_s, are found from intersection points of the dispersion curve of the propagating (waveguide or free-space) mode and that of the beam as shown in figure 3.1.

Note that in case of waveguide mode propagation, two separate solutions $f_{s1,2}$ may appear with respective group velocities v_g of the mode, which are lower ($v_{g_1} < v_{z0}$) or higher ($v_{g_2} > v_{z0}$) than the velocity v_{z0} of the electrons. In the grazing limit (beam velocity tangent to waveguide mode dispersion line) $v_g = v_{z0}$, the two solutions degenerate into a single one $f_{s_1} = f_{s_2}$.

To derive the signal of the radiation field in the time domain, the inverse Fourier transformation (3.10) is taken. In the case of two well-separated solutions, where the respective frequency bandwidths of the emission is smaller than the spectral range between the resonance frequencies, it is sufficient to use a first order approximation $k_{zq}(f) \simeq k_{zq}(f_s) + 2\pi(f - f_s)/v_g$ of the dispersion relation of the propagating mode q and calculate from the inverse Fourier transform (3.10). Assuming $t_{0i} = 0$ we obtained the field in the time

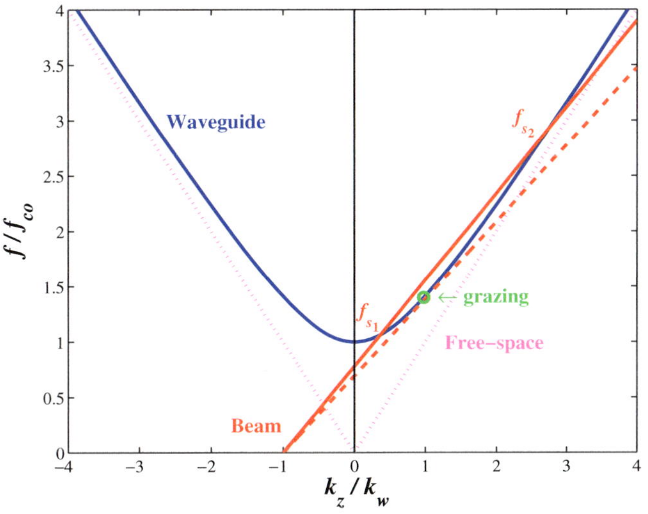

Figure 3.1: Dispersion curves of free-space, waveguide and beam modes in a FEL.

domain:

$$\mathbf{E}_q(x,y,z=L_w,t) = \Re\left\{\sum_{s=1,2} \frac{\mathcal{A}_q^{sp}}{\tau_{sp_s}} \mathrm{rect}\left(\frac{t-t_{d_s}}{\tau_{sp_s}}\right) \widetilde{\mathcal{E}}_q(x,y) e^{+j[2\pi f_s t - k_{zq}(f_s)L_w]}\right\} \tag{3.81}$$

The wave-packet emitted from the a single particle in a waveguide is observed to consist of two rectangular pulses with temporal duration (*slippage time*) $\tau_{sp_{1,2}} = L_w/v_{z0} - L_w/v_{g_{1,2}}$ modulating carriers at the synchronism frequencies $\omega_{1,2}$ respectively. Note that the radiation pulse at the upper frequency propagates faster than the radiating particle, while the low frequency pulse lags behind it. $t_{d_{1,2}} = \frac{1}{2}\left(L_w/v_{z0} + L_w/v_{g_{1,2}}\right)$ is the time delay of the center of the pulses. The time dependent signal of the radiation field emitted by a single particle passing through a wiggler is shown in figure 3.2.

The spectral density of the radiation energy emitted at the point $z = L_w$ by a single electron is calculated from (3.80) according to (3.68):

$$\frac{\mathrm{d}\,\mathcal{W}_q(L_w)}{\mathrm{d}f} = \frac{1}{2}\left|\mathcal{A}_q^{sp}\right|^2 \mathcal{N}_q \,\mathrm{sinc}^2\left(\frac{1}{2}\theta_q L_w\right) \tag{3.82}$$

A curve of the spectral energy distribution of the radiation emitted by a single electron when passing through an undulator is given in Fig. 3.3. The typical

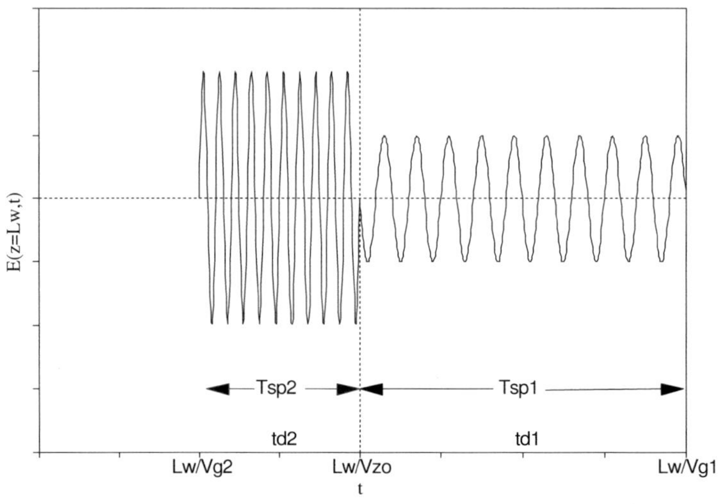

Figure 3.2: Radiation wave-packet emitted by a single wiggling particle.

bandwidth of the main lobe of the energy spectral distribution function is approximately $\Delta f_{1,2} \simeq \frac{1}{\tau_{sp_{1,2}}}$ when the resonance frequencies f_{s_1} and f_{s_2} are well separated. In the case of grazing, the bandwidth is wider, resulting in a short pulse in the time domain. The total energy contained in the wave-packed of radiation carried by mode q is found after integration of Eq. (3.82) over the entire frequency (detuning) range, results in $\mathcal{W}_q^{sp} = \frac{1}{2\,\tau_{sp_{1,2}}} \left|\mathcal{A}_q^{sp}\right|^2 \mathcal{N}_q$.

The total radiation field emitted when a stream of electrons passes through the wiggler is a summation of the fields radiated by the individual particles. When a continuous unmodulated beam having no velocity spread (*'cold' beam*) is considered, each electron arrives at a different time t_{0i} and enters the interaction region with a random phase $\phi_i = 2\pi f t_{0i}$. The fields radiated from different electrons add up incoherently, resulting in a 'shot noise', which can also be regarded as caused by fluctuations in the current due to the discreteness of the radiating particles. The average rate of emissions in an electron beam of average (DC) current I_0 is I_0/e, and the spectral density of the shot noise power carried by the qth transverse mode is given by:

$$\frac{\mathrm{d}\,P_q(L_w)}{\mathrm{d}f} = \frac{I_0}{2e} \left|\mathcal{A}_q^{sp}\right|^2 \mathcal{N}_q \,\mathrm{sinc}^2\left(\frac{1}{2}\theta_q L_w\right) \tag{3.83}$$

The total spontaneous emission power carried by the qth transverse mode is

$P_q(L_w) = \frac{1}{2\,\tau_{sp}} \frac{I_0}{e} \left|\mathcal{A}_q^{sp}\right|^2 \mathcal{N}_q.$

3.8.3 Super-radiant emission from a single bunch of electrons

When the electrons are inserted in a short bunch, which its temporal duration is shorter than the oscillation period of the undulator synchrotron radiation emitted by each of them, the radiation wave-packets combines coherently resulting in a super-radiant emission.

The current density is a product of the space-charge density and the velocity $\mathbf{v}$ of the electrons:

$$\mathbf{J}(\mathbf{r}, t) = -Q\;\mathbf{v}\;p(x, y)\;\delta\left[z - z_b(t)\right] \tag{3.84}$$

where $p(x, y)$ is the transverse space-charge distribution of the bunch of total charge Q. In the space-frequency domain the current density of the single bunch is the Fourier transform of (3.84):

$$\mathbf{J}(\mathbf{r}, f) = \int_{-\infty}^{+\infty} \mathbf{J}(\mathbf{r}, t)\; e^{-j2\pi f t}\; \mathrm{d}t = -Q\,\frac{\mathbf{v}}{v_z}\;p(x, y)\;e^{-j2\pi f t_b(z)} \tag{3.85}$$

where

$$t_b(z) = t_0 + \int_0^z \frac{\mathrm{d}z'}{v_z(z')} \tag{3.86}$$

is the time when the bunch, which entered the interaction region at a time t_0 and moved at an instantaneous axial velocity $v_z(z)$, arrives to a point z. Substitution of the expression for $\widetilde{\mathbf{J}}(\mathbf{r}, f)$ into excitation equations (3.64), (3.65) and keeping only the phase-matched components (similar to the procedure used in section 3.8.1), we find the excitation equation for the super-radiant emission:

$$\begin{aligned} &\frac{\mathrm{d}}{\mathrm{d}z}\widetilde{C}_{\pm q}(z, f) = \\ &= \pm\frac{Q\zeta_q}{2v_z\mathcal{N}_q}\; e^{-j\left[2\pi f t_0 \mp \int_0^z \theta_q(z')dz'\right]} \int\!\!\int p(x, y)\;\widetilde{\mathcal{V}}_\perp^w \cdot \widetilde{\mathcal{E}}_{\pm q}^*(x, y)\;\mathrm{d}x\,\mathrm{d}y \end{aligned} \tag{3.87}$$

Under the assumption that the effect of the radiation on the electrons in the bunch is negligible, the solution of the last equation (3.87) at the exit of a wiggler of length L_w is found to be:

$$\widetilde{C}_{+q}(L_w, f) = \mathcal{A}_q \ \mathrm{sinc}\left(\frac{1}{2}\theta_q L_w\right) \ e^{-j\left(2\pi f t_0 + \frac{1}{2}\theta_q L_w\right)} \tag{3.88}$$

where

$$\mathcal{A}_q = \frac{Q\zeta_q}{2\mathcal{N}_q} \frac{L_w}{v_{z0}} \int\!\!\int p(x,y) \ \widetilde{\mathcal{V}}_\perp^w \cdot \widetilde{\mathcal{E}}_{q\perp}^*(x,y) \ \mathrm{d}x \ \mathrm{d}y \tag{3.89}$$

Note that the super-radiant emission of a single short bunch is a coherent summation of the radiation emitted by all the electrons in the bunch.

The radiation field in the time domain, is found by inverse Fourier transformation according to (3.10):

$$\begin{aligned} &\mathbf{E}_q(x, y, z = L_w, t) = \\ &= \Re\left\{ \sum_{s=1,2} \frac{\mathcal{A}_q}{\tau_{sp_s}} \ \mathrm{rect}\left[\frac{t - (t_0 + t_{d_s})}{\tau_{sp_s}}\right] \ \widetilde{\mathcal{E}}_q(x,y) \ e^{+j[2\pi f_s t - k_{zq}(f_s) L_w]} \right\} \end{aligned} \tag{3.90}$$

The two wave-packets consist of rectangular pulses modulating carriers at frequencies $f_{1,2}$. The temporal duration of each wave-packet is the slippage time $\tau_{sp_{1,2}} = L_w/v_{z0} - L_w/v_{g_{1,2}}$, and $t_{d_{1,2}} = \frac{1}{2}\left(L_w/v_{z0} + L_w/v_{g_{1,2}}\right)$ is the time delay of the center of the wave-packets (see figure 3.2).

The spectral density of the radiation energy emitted by the electron bunch is calculated from (3.88) according to (3.68) for each of the two frequencies:

$$\frac{\mathrm{d}\mathcal{W}_q(L_w)}{\mathrm{d}f} = \frac{1}{2} \left|\mathcal{A}_q\right|^2 \mathcal{N}_q \ \mathrm{sinc}^2\left(\frac{1}{2}\theta_q L_w\right) \tag{3.91}$$

The total radiation energy contained in each wave-packet is $\mathcal{W}_q = \frac{\left|\mathcal{A}_q\right|^2 \mathcal{N}_q}{2\,\tau_{sp_{1,2}}}$.

3.9 Conclusion

Excitation and propagation of electromagnetic fields in a distributed medium with a uniform cross-section is described in the framework of a space-frequency approach. Transformation of Maxwell's equations to the frequency domain enables the presentation of the electromagnetic fields as a superposition of the eigenmodes of the "cold" medium, free of currents and charges. This

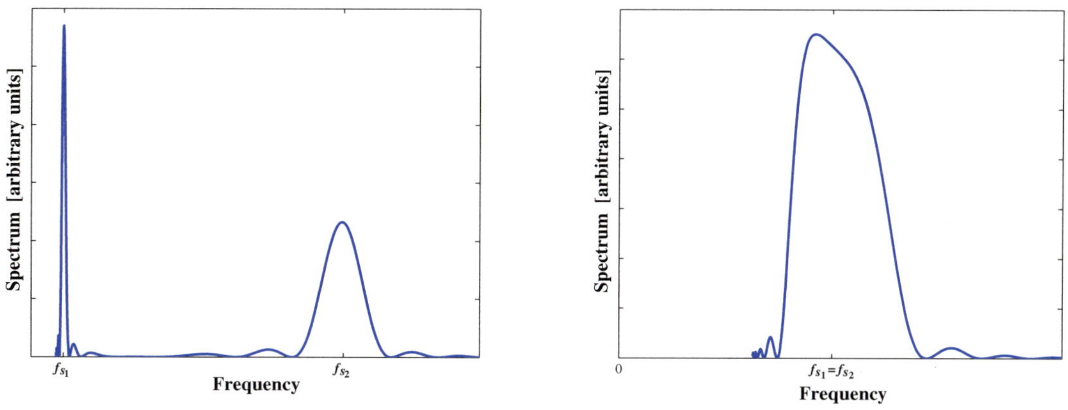

Figure 3.3: Line shape of spontaneous (Eq. (3.82)) and superradiant (Eq. (3.91)) emissions in the frequency domain. The regular case of two different resonance frequencies f_{s_1} and f_{s_2} (left), and the grazing $f_{s_1} = f_{s_2}$ (right).

coupled-mode theory enables consideration of field excitation processes and interactions with charged particles in free-electron devices in linear and non-linear regimes of operation, taking into account dispersive effects arising in such devices due to cavities, and statistical features of the electron beam.

In the approach, electromagnetic field is defined by two fast-oscillating complex functions $\tilde{V}_q(z)$ and $\tilde{I}_q(z)$, or by two slowly-oscillating complex functions $C_{+q}(z)$ and $C_{-q}(z)$. Variational implementations based on a similar approach will be discussed in the next chapter, and it will be shown that in some cases electromagnetic field can be described by a single complex function.

Bibliography

[1] J. C. Maxwell , "A dynamical theory of the electromagnetic field ," *Philosophical Transactions of the Royal Society of London*, vol. 155, 459-512, 1865.

[2] N. Wiener , "Generalized harmonic analysis," *Acta Math.*, vol. 55, 117-258, 1930.

[3] I. Khintchine, "Korrelationstheorie der stationären stochastischen Prozesse," *Mathematische Annalen*, vol. 109, no. 1, 604-615, 1934.

[4] J. H. Poynting, "On the Transfer of Energy in the Electromagnetic Field," *Philosophical Transactions of the Royal Society of London*, vol. 175, 343-361, 1884.

[5] N. Marcuvitz, and J. Schwinger, "On the Representation of the Electric and Magnetic Fields Produced by Currents and Discontinuities in Wave Guides," *J. Appl. Phys.*, vol. 22, 806-819, 1951.

[6] Y. Pinhasi, Yu. Lurie, and A. Yahalom, "Model and simulation of wide-band interaction in free-electron lasers," *Nucl. Instr. and Meth. in Phys. Res. A*, vol. 475, no. 1-3, 147-152, 2001.

[7] Y. Pinhasi, Yu. Lurie, A. Yahalom, and A. Abramovich , "Space-frequency model of amplified spontaneous emission and super-radiance in free electron laser operating in the linear and non-linear regimes," *Nucl. Instr. Meth. Phys. Res. A*, vol. 483, Issues 1-2, 510-515, 2002.

[8] Y. Pinhasi, and Yu. Lurie, "Generalized theory and simulation of spontaneous and super-radiant emissions in electron devices and free-electron lasers," *Phys. Rev. E*, vol. 65, Issue 2, 026501 [8 pages], 2002.

[9] Y. Pinhasi, Yu. Lurie, and A. Yahalom, "Space-frequency model of ultra wide-band interactions in millimeter wave masers," NATO Science Series II: Mathematics, Physics and Chemistry, Vol. 203, Springer, 2005 (Edited by J.L. Hirshfield and M.I. Petelin). Proc. of NATO advance workshop on Quasi-optical control of intense microwave transmission, Nizhny Novgorod: Russia, 2004.

[10] Y. Pinhasi, Yu. Lurie, and A. Yahalom, "Space-frequency model of ultra wide-band interactions in free-electron lasers," *Phys. Rev. E*, vol. 71, 036503 [8 pages], 2005.

[11] Y. Pinhasi, A. Yahalom, Yu. Lurie, and G.A. Pinhasi, "Backward Wave Excitation and Generation of Oscillations in Free-Electron Lasers in the Absence of Feedback — Beyond the High Gain Approximation," *IEEE Journal of Quantum Electronics*, vol. 43, no. 10, 849-854, 2007.

[12] Yu. Lurie, and Y. Pinhasi, "Enhanced super-radiance from energy-modulated short electron bunch free-electron lasers," *Phys. Rev. ST Accel. Beams*, vol. 10, 080703 [8 pages], 2007.

Advances in Classical Field Theory, 2011, 86-101

CHAPTER 4

Variational analysis of electromagnetic fields in closed and open structures

Abstract

Possible variational principles for excitation of an electromagnetic field in a wave guide are discussed. Our emphasis is not on the calculation of the modal shapes, which is common in previous art, but rather on the calculation of modal amplitude evolution, which are important in electron devices such as free electron lasers and gyrotrons. The variational formalism allows us to show that the electromagnetic field can be defined by a single complex function.

4.1 Introduction

Variational principles for electromagnetic field dynamics, including their interaction with matter are abundant in the literature [1]- [9]. Moreover, the behavior of the electromagnetic field inside a wave guide in terms of a variational principle was studied in many texts [4]- [9], most of the times in order to provide a basis for a numerical scheme. These works are concerned mainly with the derivations of eigenmodes for the case of non trivial geometries or an inhomogeneous refraction index. In this chapter we are not concerned with the modal form rather we assume that it is known. Our main concern

Asher Yahalom (Ed)

is the development of the modal amplitude inside the wave guide due to its interaction with propagating charge.

An additional difference between this chapter and previous art is the methodology we use for developing our variational principle. While previous workers initiated the development of their frequency domain variational principle from the relevant equations (usually defining the square of the equations as the variational functional). We start from the general "canonical" form of the variational principle for electromagnetic fields [10] which is stated in terms of gauge fields, we than adapt it to a form which is relevant for the physical scenario taking place inside a wave-guide. In a sense this a combination of the suggested formalisms of the previous two chapters.

The structure of this chapter is as follows: First we discuss the the vector potential modal representation. After that we review the classical variational principle of the electromagnetic field and its variational derivatives. Next the action is represented in the spectral - modal scheme and second order equations are obtained for the fields amplitudes. Finally the introduction of the quasi Hamiltonian allows us to obtain first order equations in terms of the field amplitudes, while the introduction of backward-forward waves puts the variational principle in a particular simple form which concludes our chapter.

4.2 Modal presentation of the electromagnetic field in the frequency domain using vector potentials

This section presents the formalism employed in the previous chapter for analysing the excitation of electromagnetic fields by current sources distributed along a wave guide [11, 12, 13], but now in the language of vector potentials. The approach taken here utilizes representation of the total electromagnetic fields and their sources in terms of vector functions, which are the eigenmodes solutions of the medium, free of charge or current sources. The 'phasor like' quantities defined in (3.6) can be expanded in terms of transverse eigenmodes of the medium in which the field is excited and propagates. The perpendicular component of the electric and magnetic fields are given in any cross-section as a linear superposition of a complete set of transverse eigenmodes:

$$\begin{aligned}\widetilde{\mathbf{E}}_{\perp}(\mathbf{r},\omega) &= \sum_q V_q(z,\omega)\widetilde{\mathcal{E}}_{q\perp}(x,y)\\ \widetilde{\mathbf{H}}_{\perp}(\mathbf{r},\omega) &= \sum_q I_q(z,\omega)\widetilde{\mathcal{H}}_{q\perp}(x,y)\end{aligned}\tag{4.1}$$

The summations include propagating and cut-off TE and TM modes, for which $V_q(z,\omega)$ and $I_q(z,\omega)$ are the scalar amplitude of the electric and magnetic fields respectively and $\widetilde{\mathcal{E}}_{q\perp}(x,y)$ and $\widetilde{\mathcal{H}}_{q\perp}(x,y)$ are their respective profiles.

Expressions for the longitudinal component of the electric and magnetic fields are obtained after substituting the modal representation (4.1) of the fields into Maxwell's equations, where the Fourier transform of the current density $\mathbf{J}$, $\widetilde{\mathbf{J}}(\mathbf{r},\omega)$ is introduced[1]:

$$\begin{aligned}\widetilde{E}_z(\mathbf{r},\omega) &= \sum_q I_q(z,\omega)\widetilde{\mathcal{E}}_{qz}(x,y) + \frac{1}{j\omega\epsilon}\widetilde{J}_z(\mathbf{r},\omega)\\ \widetilde{H}_z(\mathbf{r},\omega) &= \sum_q V_q(z,\omega)\widetilde{\mathcal{H}}_{qz}(x,y)\end{aligned}\tag{4.2}$$

By imposing the appropriate boundary conditions, the Maxwell vector equations are transformed into scalar differential ('transmission line') equations, which describe the evolution of the equivalent electric and magnetic amplitudes $V_q(z,\omega)$ and $I_q(z,\omega)$:

$$\begin{aligned}-\frac{dV_q(z,\omega)}{dz} &= -jk_{zq}I_q(z,\omega) + v_q(z,\omega)\\ -\frac{dI_q(z,\omega)}{dz} &= -jk_{zq}V_q(z,\omega) + i_q(z,\omega)\end{aligned}\tag{4.3}$$

where:

$$k_{zq} = \begin{cases} j\sqrt{k_{\perp q}^2 - k^2} = j|k_{zq}| & k < k_{\perp q} \quad \text{(cut-off modes)}\\ \sqrt{k^2 - k_{\perp q}^2} = |k_{zq}| & k > k_{\perp q} \quad \text{(propagating modes)}\end{cases}\tag{4.4}$$

[1] The non physical magnetic current density is ignored in this chapter

is the axial wave number of mode q and:

$$\begin{aligned} v_q(z,\omega) &\equiv \frac{1}{\mathcal{N}_q}\int\int \widetilde{J}_z\widetilde{\mathcal{E}}^*_{qz}dxdy \\ i_q(z,\omega) &\equiv \frac{1}{\mathcal{N}_q}\frac{Z_q}{Z_q^*}\int\int \widetilde{\mathbf{J}}_\perp\cdot\widetilde{\mathcal{E}}^*_{q\perp}dxdy \end{aligned} \tag{4.5}$$

The normalization of the field amplitudes of each mode is made via each mode's complex Poynting vector power:

$$\mathcal{N}_q = \int\int_{c.s.}\left[\widetilde{\mathcal{E}}_{q\perp}(x,y)\times\widetilde{\mathcal{H}}^*_{q\perp}(x,y)\right]\cdot\hat{\mathbf{z}}dxdy \tag{4.6}$$

and the mode impedance is given by:

$$Z_q = \begin{cases} \sqrt{\frac{\mu}{\epsilon}}\frac{k}{k_{zq}} = \frac{\omega\mu}{k_{zq}} & \text{for TE modes} \\ \sqrt{\frac{\mu}{\epsilon}}\frac{k_{zq}}{k} = \frac{k_{zq}}{\omega\epsilon} & \text{for TM modes} \end{cases} \tag{4.7}$$

ε is the electric susceptibility and μ is the magnetic permeability.

The transmission-line equations (4.3) can also be written in the form:

$$\begin{aligned} V_q''(z,\omega) + k_{zq}^2V_q(z,\omega) &= -v_q'(z,\omega) - jk_{zq}i_q(z,\omega) \\ I_q''(z,\omega) + k_{zq}^2I_q(z,\omega) &= -jk_{zq}v_q(z,\omega) - i_q'(z,\omega) \end{aligned} \tag{4.8}$$

where ($'$) denotes a derivative in respect to z. Notice that only one of the equations in (4.8) needs to be solved, since solving for $V_q(z,\omega)$ we obtain immediately the solution for $I_q(z,\omega)$ through equation (4.3) .

The scalar potential $\Phi(\mathbf{r},t)$ and vector potential $\mathbf{A}(\mathbf{r},t)$ are related to the electric and magnetic fields by:

$$\begin{aligned} \mu\mathbf{H}(\mathbf{r},t) &= \nabla\times\mathbf{A}(\mathbf{r},t) \\ \mathbf{E}(\mathbf{r},t) &= -\frac{\partial}{\partial t}\mathbf{A}(\mathbf{r},t) - \nabla\Phi(\mathbf{r},t) \end{aligned} \tag{4.9}$$

Choosing a gauge transformation where $\Phi(\mathbf{r},t) = 0$, the vector potential is given by the time integration:

$$\mathbf{A}(\mathbf{r},t) = -\int_{-\infty}^{t}\mathbf{E}(\mathbf{r},t')dt' \tag{4.10}$$

Consequently, in the frequency domain:

$$\widetilde{\mathbf{A}}(\mathbf{r},\omega) = \frac{1}{j\omega}\widetilde{\mathbf{E}}(\mathbf{r},\omega) \tag{4.11}$$

Using expansion (4.1), the perpendicular component of the vector potential can be presented by:

$$\widetilde{\mathbf{A}}_\perp(\mathbf{r},\omega) = \frac{1}{j\omega}\sum_q V_q(z,\omega)\widetilde{\mathcal{E}}_{q\perp}(x,y) \tag{4.12}$$

and from (4.2), the longitudinal component of the vector potential is:

$$\widetilde{A}_z(\mathbf{r},\omega) = \frac{1}{j\omega}(\sum_q I_q(z,\omega)\widetilde{\mathcal{E}}_{qz}(x,y) + \frac{1}{j\omega\epsilon}\widetilde{J}_z(\mathbf{r},\omega)) \tag{4.13}$$

In terms of equation (4.5) the longitudinal component can be shown to have the form:

$$\widetilde{A}_z(\mathbf{r},\omega) = \frac{1}{j\omega}\sum_q (I_q^{TM}(z,\omega) - \frac{jk_{z_q}}{k_{\perp_q}^2}v_q^{TM}(z,\omega))\widetilde{\mathcal{E}}_{q_z}(x,y) \tag{4.14}$$

In this case equation (4.3) serve as a definition for I_q :

$$I_q \equiv \frac{\partial_z V_q + v_q}{jk_{z_q}}. \tag{4.15}$$

4.3 The classical Variational Principle

It is well known that the action of the electromagnetic field can be given in terms of the vector potential in the following form [10]:

$$\mathcal{A} = \int [\frac{1}{2}(\varepsilon(\partial_t \mathbf{A})^2 - \frac{1}{\mu}(\nabla \times \mathbf{A})^2) + \mathbf{J}\cdot\mathbf{A}]d^3x dt. \tag{4.16}$$

Taking the variational derivative of $\mathcal{A}$ with respect the vector potential $\mathbf{A}$ we obtain:

$$\delta\mathcal{A} = \delta\mathcal{A}_t + \delta\mathcal{A}_{boundary} + \delta\mathcal{A}_{equations}. \tag{4.17}$$

$\delta\mathcal{A}_t$ is the time condition term given by:

$$\delta\mathcal{A}_t = \varepsilon\int \partial_t\mathbf{A}\cdot\delta\mathbf{A}d^3x|_{t_1}^{t_2} \tag{4.18}$$

t_1 is the initial time of the system and t_2 is the final time. $\delta\mathcal{A}_{boundary}$ is the boundary term:

$$\delta\mathcal{A}_{boundary} = \oint \mathbf{H} \times \delta\mathbf{A} \cdot d\vec{S}dt \tag{4.19}$$

the integral is taken over all the surface containing the volume under consideration. Finally, the equation part is given by:

$$\delta\mathcal{A}_{equations} = \int \delta\mathbf{A} \cdot (\varepsilon\partial_t\mathbf{E} - \nabla \times \mathbf{H} + \mathbf{J})d^3xdt. \tag{4.20}$$

Provided that $\delta\mathcal{A}_t$ and $\delta\mathcal{A}_{boundary}$ vanish for otherwise arbitrary variations than $\delta\mathcal{A}_{total} = 0$ vanishes if and only if:

$$\nabla \times \mathbf{H} = \varepsilon\partial_t\mathbf{E} + \mathbf{J}. \tag{4.21}$$

The other Maxwell equations are satisfied automatically by virtue of the vector potential representation equation (4.9) and by virtue of the continuity equation.

4.4 Spectral Representation

In order to use the spectral representation in the variational technique one must assure in advance that the condition $\delta\mathcal{A}_t = 0$ (equation (4.18)) is taken care of. This can be done by choosing $t_1 = 0$ and $t_2 = T$ and demanding that $\mathbf{A}(0) = \mathbf{A}(T)$. Thus $\mathbf{A}$ is given (instead of Fourier transform of the type given in equation (3.5)) by the Fourier Sum:

$$\mathbf{A} = \sum_{n=-\infty}^{\infty} \mathbf{A}_n e^{j\omega_n t}, \qquad \omega_n = \frac{2\pi n}{T} \tag{4.22}$$

for which:

$$\mathbf{A}_n = \frac{1}{T}\int_0^T \mathbf{A} e^{-j\omega_n t} dt \tag{4.23}$$

and $\mathbf{A}_{-n} = \mathbf{A}_n^*$ since $\mathbf{A}$ is real.

In terms of this notation equation (4.16) takes the form:

$$\mathcal{A} = T[\sum_{n=1}^{\infty} \mathcal{A}_n + \mathcal{A}_{DC}] \tag{4.24}$$

In which:

$$\begin{aligned}
\mathcal{A}_n &= \int [\varepsilon\omega_n^2|\mathbf{A}_n|^2 - \frac{1}{\mu}|\nabla\times\mathbf{A}_n|^2 + \mathbf{J}_n^*\cdot\mathbf{A}_n + \mathbf{J}_n\cdot\mathbf{A}_n^*]d^3x \\
&= \int [\varepsilon\omega_n^2|\mathbf{A}_n|^2 - \frac{1}{\mu}|\nabla\times\mathbf{A}_n|^2 + 2\Re(\mathbf{J}_n^*\cdot\mathbf{A}_n)]d^3x \\
\mathcal{A}_{DC} &= \int [-\frac{1}{2\mu}|\nabla\times\mathbf{A}_0|^2 + \mathbf{J}_0\cdot\mathbf{A}_0]d^3x \qquad (4.25)
\end{aligned}$$

$\mathbf{J}_n$ is the Fourier component of the current density, $\Re(x)$ stands for the real part of x, $|x|$ stands for the absolute value of x and x^* stands for the complex conjugate of x. The action given in equation (4.24) and equation (4.25) is obviously real (despite its apparent complex nature). From now on we disregard $\mathcal{A}_{DC}$ which is not relevant for time dependent source currents.

4.5 The Action in a Wave Guide

Inserting the modal representation equation (4.12) and equation (4.13) into equations (4.24,4.25) we obtain:

$$\begin{aligned}
\mathcal{A} &= \frac{T^2}{\pi}\sum_{n=1}^{\infty}\frac{1}{n}\sum_q \frac{\mathcal{N}_{q,n}^*}{k_{z_{q,n}}}L_{q,n} \\
L_{q,n} &\equiv \int \mathcal{L}_{q,n}dz \\
\mathcal{L}_{q,n} &\equiv \frac{1}{2}k_{z_{q,n}}^2|V_{q,n}|^2 - \frac{1}{2}|\partial_z V_{q,n}|^2 - \frac{1}{2}\frac{\omega_n^2}{c^2k_{\perp_q}^2}|v_{q,n}|^2 - \Re(v_{q,n}^*\partial_z V_{q,n}) \\
&- \Im(k_{z_{q,n}}i_{q,n}V_{q,n}^*) \qquad (4.26)
\end{aligned}$$

in which $\Im(x)$ stand for the imaginary part x. From appendix A (see also [13]) it is obvious that $\frac{\mathcal{N}_{q,n}^*}{k_{z_{q,n}}}$ is real and so are the Lagrangian $L_{q,n}$ and the Lagrangian density $\mathcal{L}_{q,n}$. The quantities $i_{q,n}, v_{q,n}$ are defined in equations (4.5). From now on we will suppress the indices q, n.

In terms of the amplitude V and its complex conjugate V^*, the Lagrangian density $\mathcal{L}$ can be written as:

$$\mathcal{L} = \frac{1}{2}\{k_z^2VV^* - \partial_zV\partial_zV^* - \frac{\omega^2}{c^2k_\perp^2}vv^* - \partial_zVv^* - \partial_zV^*v + jk_ziV^* - jk_z^*i^*V\} \qquad (4.27)$$

It should be noted that using the modal representation given in equation (4.12) the boundary $\delta\mathcal{A}_{boundary}$ term given in equation (4.19) vanishes on the circumference of the wave-guide in the modal representation and the only contributions come from the planes $z=0$ and $z=L_w$ which amount to the condition:

$$\Im(k_z I \delta V^*)|_0^{L_w} = 0 \tag{4.28}$$

for each mode.

Taking the variational derivative of equation (4.27) we obtain:

$$\delta L = \frac{1}{2}\{\int [jk_z i + k_z^2 V + \partial_z^2 V + \partial_z v]\delta V^* dz - (\partial_z V + v)\delta V^*|_0^{L_w} + c.c.\} \tag{4.29}$$

in which *c.c.* stands for complex conjugate. The boundary conditions given in equation (4.29) are the same as in equation (4.28) by virtue of equation (4.15) . The equations obtained are the same as in equation (4.8) and its complex conjugate.

4.6 Some Numerical Aspects

At this stage it is already clear that the Lagrangian appearing in equation (4.26) is real. However, in order to understand its mathematical structure we shall write it in terms of real quantities. Every complex number C can be written as $C = C_r + jC_i$ in which C_r, C_i are real numbers and $C_r = \Re(C)$ is the real part of C while $C_i = \Im(C)$ is the imaginary part of C. Representing all the quantities appearing in $\mathcal{L}$ (equations (4.26)) in terms of their real and imaginary parts we arrive at the result:

$$\begin{aligned}
\mathcal{L} &\equiv \frac{1}{2}k_z^2|V|^2 - \frac{1}{2}|\partial_z V|^2 - \frac{1}{2}\frac{\omega^2}{c^2 k_\perp^2}|v|^2 - \Re(v^*\partial_z V) - \Im(k_z i V^*) \\
&= \frac{1}{2}\left[k_z^2(V_r^2 + V_i^2) - (\partial_z V_r^2 + \partial_z V_i^2) - \frac{\omega^2}{c^2 k_\perp^2}(v_r^2 + v_i^2) - \partial_z V_r v_r - \partial_z V_i v_i\right] \\
&- \frac{1}{2}\begin{cases} k_z(i_i V_r - i_r V_i) & \text{(propagating modes)} \\ |k_z|(i_r V_r - i_i V_i) & \text{(cut-off modes)} \end{cases}
\end{aligned} \tag{4.30}$$

Notice that the cut-off modes Lagrangian density decouples into two separate Lagrangian densities:

$$\begin{aligned}\mathcal{L} &= \mathcal{L}_r + \mathcal{L}_i \\ \mathcal{L}_r &= \frac{1}{2}\left[k_z^2 V_r^2 - (\partial_z V_r)^2 - \frac{\omega^2}{c^2 k_\perp^2} v_r^2 - \partial_z V_r v_r - |k_z| i_r V_r\right] \\ \mathcal{L}_i &= \frac{1}{2}\left[k_z^2 V_r i^2 - (\partial_z V_i)^2 - \frac{\omega^2}{c^2 k_\perp^2} v_i^2 - \partial_z V_i v_i + |k_z| i_i V_i\right] \qquad (4.31)\end{aligned}$$

while the propagating modes Lagrangian density cannot decouple. Using any type of discritization the Lagrangian density given in equation (4.30) will become a real bilinear form. For cut-off modes the form of $-\mathcal{L}$ appears to be positive since $k_z^2 = -|k_z^2|$ according to equation (4.4) . Thus the solution will correspond to the minimum of the bilinear form which can be found by standard numerical techniques such as the conjugate gradient method [15]. For propagating modes $k_z^2 = |k_z^2|$ the solution will correspond to a saddle point of the linear form and can be found using techniques such as the ones described in [16]-[18].

4.7 The Quasi Hamiltonian

In certain cases it is desirable to obtain first order equations instead of the second order equations (4.8). In analytical mechanics [14] their is a well known technique to reach this goal using the Hamiltonian construction. Since L given in equation (4.26) is not strictly speaking a Lagrangian (time which appears in proper Lagrangians is replaced here by the longitudinal coordinate z) we will denote the analogue construction of the Hamiltonian a "quasi Hamiltonian". For convenience we introduce the Lagrangian density $\bar{\mathcal{L}}$:

$$\begin{aligned}\bar{\mathcal{L}} &= -2\mathcal{L} \\ &= \partial_z V \partial_z V^* - k_z^2 V V^* + \frac{\omega^2}{c^2 k_\perp^2} v v^* \\ &+ \partial_z V v^* + \partial_z V^* v - j k_z i V^* + j k_z^* i^* V \qquad (4.32)\end{aligned}$$

in which we utilized equations (4.27). Next we define the quasi canonical momentums of $V' \equiv \partial_z V$ and $V'^* \equiv \partial_z V^*$:

$$\Pi \equiv \frac{\partial \bar{\mathcal{L}}}{\partial V'} = V'^* + v^* = -j k_z^* I^* \qquad \Pi^* \equiv \frac{\partial \bar{\mathcal{L}}}{\partial V'^*} = V' + v = j k_z I \qquad (4.33)$$

in which equation (4.15) is used. Notice that the quasi canonical momentums are proportional to I^* and I respectively. Having done this we are in a position to define the quasi Hamiltonian density:

$$\begin{aligned}
\mathcal{H} &\equiv V'\Pi + V'^*\Pi^* - \bar{\mathcal{L}} \\
&= |k_z|^2|I|^2 + k_z^2|V|^2 - jk_zIv^* + jk_z^*I^*v - \frac{k_z^2}{k_\perp^2}|v|^2 + jk_ziV^* - jk_z^*i^*V \\
&= k_z\left[k_z(|V|^2 \pm |I|^2) - jIv^* \pm jI^*v - \frac{k_z}{k_\perp^2}|v|^2 + jiV^* \mp ji^*V\right] \qquad (4.34)
\end{aligned}$$

the upper sign should be attributed to propagating modes while the lower signs should be attributed to decaying modes. Thus $\bar{\mathcal{L}}$ can be written as:

$$\begin{aligned}
\bar{\mathcal{L}} &= V'\Pi + V'^*\Pi^* - \mathcal{H} \\
&= k_z[\mp jI^*V' + jIV'^* - k_z(|V|^2 \pm |I|^2) + jIv^* \mp jI^*v - jiV^* \pm ji^*V \\
&+ \frac{k_z}{k_\perp^2}|v|^2] \qquad (4.35)
\end{aligned}$$

Our next step will be to take the variational derivative with respect to I and V and their complex conjugates of $\bar{L}$ which is defined as:

$$\bar{L} = \int \bar{\mathcal{L}}dz = -2L \qquad (4.36)$$

This will lead to the expression:

$$\begin{aligned}
\delta\bar{L} &= k_z\int dz[\delta I(jV'^* \mp k_zI^* + jv^*) \mp \delta I^*(jV' + k_zI + jv) \\
&+ \delta V(\pm jI'^* - k_zV^* \pm ji^*) - \delta V^*(jI' + k_zV + ji)] \\
&+ jk_z(I\delta V^* \mp I^*\delta V)|_0^{L_w} \qquad (4.37)
\end{aligned}$$

The boundary conditions term given in equations (4.37) is the same as in equation (4.29) by virtue of equations (4.15). The equations obtained are the same as equations (4.3) and their complex conjugates.

4.8 The Forward-Backward Formulation

In terms of V and I one can define the following new variables [13]:

$$C_+ \equiv \frac{1}{2}(V + I)e^{-jk_zz}, \qquad C_- \equiv \frac{1}{2}(V - I)e^{jk_zz} \qquad (4.38)$$

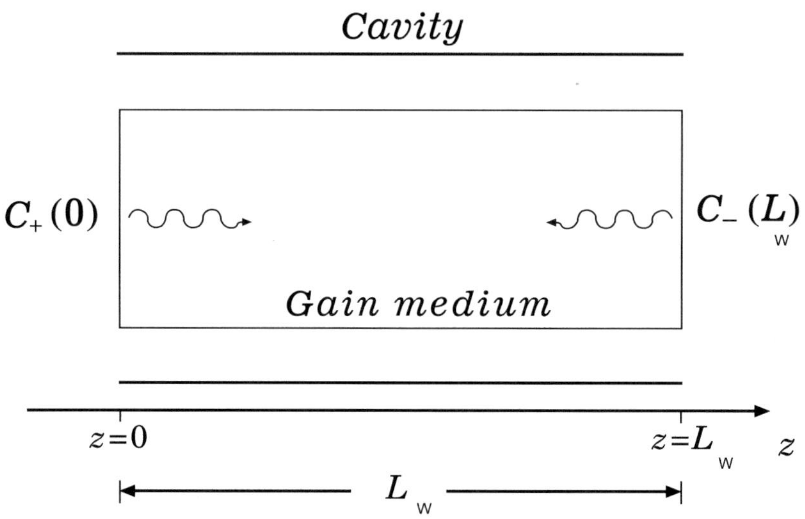

Figure 4.1: Interaction of the electromagnetic field in a gain medium

Or vice-versa as:

$$V = C_+ e^{jk_z z} + C_- e^{-jk_z z}, \qquad I = C_+ e^{jk_z z} - C_- e^{-jk_z z}. \tag{4.39}$$

Thus C_+ and C_- appear as the amplitudes of forward and backward waves respectfully (see figure 4.1) in the case of propagating modes. Inserting the above variables into $\bar{L}$ given in equation (4.36) we obtain:

$$\begin{aligned} \bar{L} &= jk_z\{2\int dz[C_-^* C_-' + C_+'^* C_+ - C_+^*\beta + C_+\beta^* \\ &+ C_-^*\alpha - C_-\alpha^* - j\frac{k_z}{2k_\perp^2}|v|^2] \\ &+(C_-^* C_+ e^{2jk_z z} - C_- C_+^* e^{-2jk_z z} - |C_-|^2 - |C_+|^2)|_0^{L_w}\} \end{aligned} \tag{4.40}$$

In which:

$$\alpha = \frac{1}{2}(v - i)e^{jk_z z} \qquad \beta = \frac{1}{2}(v + i)e^{-jk_z z}. \tag{4.41}$$

At this stage one is tempted to discard the boundary term in the above equation since it appears to have no effect on the resulting equations, however, this will lead to unphysical boundary conditions and thus should be avoided. Taking the variational derivative we obtain:

$$\delta\bar{L} = \delta\bar{L}_{equation} + \delta\bar{L}_{boundary} \tag{4.42}$$

in which:

$$\delta \bar{L}_{equation} = 2jk_z \int [(\alpha + C'_-)\delta C^*_- - (\beta + C'_+)\delta C^*_+ - c.c.]dz \tag{4.43}$$

in which $c.c.$ stand for complex conjugate and:

$$\delta \bar{L}_{boundary} = -jk_z[\delta C^*_+(C_- e^{-2jk_z z} - C_+) - \delta C^*_-(C_+ e^{2jk_z z} - C_-) - c.c.]|_0^{L_w}. \tag{4.44}$$

The boundary term given in equation (4.44) are the same as in equation (4.28) by virtue of equations (4.39). The equations obtained are:

$$C'_- = -\alpha, \qquad C'_+ = -\beta \tag{4.45}$$

and their complex conjugates which provides a truly elegant way to compute the field dynamics.

For decaying modes $\bar{L}$ given in equation (4.36) takes the form:

$$\begin{aligned} \bar{L} &= jk_z\{2\int dz[C^*_+ C'_- + C_+ C'^*_- + C^*_+\alpha + C_+\alpha^* \\ &- C^*_-\beta - C_-\beta^* - j\frac{k_z}{2k_\perp^2}|v|^2] \\ &- [|C_-|^2 e^{-2jk_z z} - |C_+|^2 e^{2jk_z z} + C_- C^*_+ + C^*_- C_+]|_0^{L_w}\} \end{aligned} \tag{4.46}$$

At this stage one is tempted to discard the boundary term in the above equation since it appears to have no effect on the resulting equations, however, this will lead to unphysical boundary conditions and thus should be avoided. Taking the variational derivative we obtain:

$$\delta \bar{L} = \delta \bar{L}_{equation} + \delta \bar{L}_{boundary} \tag{4.47}$$

in which:

$$\delta \bar{L}_{equation} = 2jk_z \int [(\alpha + C'_-)\delta C^*_+ - (\beta + C'_+)\delta C^*_- + c.c.]dz \tag{4.48}$$

in which $c.c.$ stand for complex conjugate and:

$$\delta \bar{L}_{boundary} = -jk_z[\delta C^*_+(C_- - C_+ e^{2jk_z z}) + \delta C^*_-(C_- e^{-2jk_z z} - C_+) + c.c.]|_0^{L_w}. \tag{4.49}$$

The boundary term given in equation (4.49) are the same as in equation (4.28) by virtue of equations (4.39). The equations obtained are the same as in equation (4.45) and their complex conjugates which provides a truly elegant way to compute the field dynamics.

4.9 Conclusions

Three different action principles were obtained in this work: one in terms of the V modal amplitude leading to second order equations. Another principle was formulated in terms of the V, I amplitudes through the quasi Hamiltonian concept leading to first order equations. And finally an action principle in terms of the forward and backward modes were derived including the correct boundary conditions for those equations. The action can be used as a basis for a numerical scheme as outlined in section 4.6. It was observed that different numerical techniques should be used for propagating and cut-off modes. Additional possible applications of the above variational principles include derivation of constants of motion using the Noether theorem and quantization of the electromagnetic field in a wave-guide using the action in a path integral technique. The result in a further advancement in the theory of the the electromagnetic field enabling the derivation of all properties of the field from a single complex function V.

4.10 Appendix: The complex nature of the electromagnetic profiles in the wave guide

Table 4.1: Parameters of propagating and cut-off modes

Mode type:		Propagating	Cut-off
Axial wavenumber	k_{zq}	real	imaginary
Impedance:	Z_q	real	imaginary
<u>*TE* mode</u>:			
Longitudinal magnetic field component:	$\widetilde{\mathcal{H}}_{qz}$	real	real
Transverse magnetic field component:	$\widetilde{\mathcal{H}}_{q\perp}$	imaginary	real
Transverse electric field component:	$\widetilde{\mathcal{E}}_{q\perp}$	imaginary	imaginary
<u>*TM* mode</u>:			
Longitudinal electric field component:	$\widetilde{\mathcal{E}}_{qz}$	real	real
Transverse electric field component:	$\widetilde{\mathcal{E}}_{q\perp}$	imaginary	real
Transverse magnetic field component:	$\widetilde{\mathcal{H}}_{q\perp}$	imaginary	imaginary
Power normalization	$\mathcal{N}_q$	real	imaginary

Bibliography

[1] T. Fla, " Select this Article Action principle and the Hamiltonian formulation for the MaxwellVlasov equations on a symplectic leaf", *Physics of Plasmas*, vol. 1, issue 8, 2409-2418, 1994.

[2] R.L. Seliger, and G.B. Whitham, "Variational Principles in Continuum Mechanics," *Proc. Roy. Soc. London A*, vol. 305, 1, 1968.

[3] R. Elvsen, and J. Larsson, "An action principle for the relativistic VlasovMaxwell system," *Physica Scripta*, vol. 47, issue 4, 571-575, 1993.

[4] I.V. Lindell, "Variational methods for nonstandard eigenvalue problems in waveguide and resonator analysis," *IEEE Transactions on Microwave Theory and Techniques*, vol. 30, no. 8, 1194-1204, August 1982.

[5] B.E. Little, and H.A. Haus, "A variational coupled-mode theory for periodic waveguides," *IEEE Journal of Quantum Electronics*, vol.31, issue 12 ,2258-2264 ,1995.

[6] G. Berry, S. Burke, C. Smartt, T. Benson, and P. Kendel, "Exact and variational Fourier transform methods for analysis of multilayered planar waveguides," *IEE Proceedings-Optoelectronics*, vol. 142, issue 1, 66-75, 1995.

[7] S.S. Patrick, and K.J. Webb, "A variational vector finite difference analysis for dielectric waveguides," *IEEE Transactions on Microwave Theory and Techniques*, vol. 40, issue 4, 692-698, 1992.

[8] A.N. Bogolyubov, A.L. Delitsyn, and I.E. Mogilevskii, "Variational finite-difference method of waveguide-system modeling and spectral problems of waveguide theory," *Journal of Communications Technology and Electronics*, vol. 45, supp 2, s126-s130, 2000.

[9] I.A. Erteza and J.W. Goodman, "A variational derivation of the transverse vector wave equation for application to dielectric waveguide analysis," *Optics Communications*, vol. 115, issue 1-2, 52-56, 1995.

[10] J.D.Jackson, Classical Electrodynamics, Third Edition. Wiley: New York, 1999.

[11] N. Markuvitz, and J. Schwinger, "On the representation of electric and magnetic fields produced by currents and discontinuities in waveguides," *J. Appl. Phys.*, vol. 22, 806, 1951.

[12] R.E. Collin, Foundations for Microwave Engineering. Mcgraw-Hill: Singapore, 1966.

[13] Y. Pinhasi, "Analysis of Super-Radiance Effects and of Transverse Mode Excitation and Coupling in Free-Electron Lasers", PhD Thesis, Tel-Aviv University, Israel, 1995.

[14] H. Goldstein, Classical Mechanics. Addison-Wesley: Reading MA., 1950.

[15] W.H. Press, S.A. Teukolsky, W.T. Vetterling, and B.P. Flannery, Numerical Recipes in C The Art of Scientific Computing, Second Edition. Cambridge University Press, 1992.

[16] M.R. Cui, "Analysis of iterative algorithms of Uzawa type for saddle point problems," *Applied Numerical Mathematics*, vol. 50, issue 2 , 133-146, August 2004.

[17] P. Ciarlet Jr., J. Huang, and J. Zou, "Some observations on generalized saddle-point problems," *SIAM Journal on Matrix Analysis and Applications*, vol. 25, no. 1, 224-236, 2003.

[18] J.T. Hoag, "Monotonicity of solutions converging to a saddle point equilibrium ," *Journal of Mathematical Analysis and Applications*, vol. 295, issue 1 , 10-14, July 2004.

CHAPTER 5

Advances in the field theory of magnetohydrodynamics

Abstract

It will be shown that barotropic magnetohydrodynamics is equivalent to a four function field theory, reducing the number of equations and variables needed to describe the theory from seven (the magnetic field $\vec{B}$ the velocity field $\vec{v}$ and the density ρ) to four-function field theory possesses a novel double infinite symmetry group. Non-trivial topologies and the differences between stationary and non-stationary flows will be given detailed attention.

5.1 Introduction to Variational Principles of Magnetohydrodynamics

The theory of magnetohydrodynamics is in a sense a combination of field theories discussed in the previous chapters, since is involves an interaction of the magnetic field with the flow field. As in previous chapters we will attempt to reduce the number of functions involved in the field theory of Magnetohydrodynamics. In order to achieve this we will need to go through the already familiar route of Variational analysis. Variational principles for magnetohydrodynamics were introduced by previous authors both in Lagrangian and Eulerian form. Following the work of Eckart [1] for non-magnetic flows, Newcomb [2] has introduced in his paper a Lagrangian variational formalism

Asher Yahalom (Ed)

for magnetohydrodynamics. A similar formalism was discussed in Sturrock's book [3]. Eulerian variational principles for non-magnetic fluid dynamics were first introduced by Davydov [4] as was previously discussed. Following the work of Davydov, Zakharov and Kuznetsov [5] suggested an Eulerian variational principle for magnetohydrodynamics. However, the variational principle suggested by Zakharov and Kuznetsov contained **two** more functions than the standard formulation of magnetohydrodynamics with a total sum of **nine** variational variables. Another Eulerian variational principle for magnetohydrodynamics was introduced independently by Calkin [6] in a work that preceded Zakharov and Kuznetsov paper by seven years. However, Calkin's variational principle also depends on as much as eleven variational variables. The situation was somewhat improved when Vladimirov and Moffatt [7] in a series of papers have discussed an Eulerian variational principle for incompressible magnetohydrodynamics. Their variational principle contained only three more functions in addition to the seven variables which appear in the standard equations of magnetohydrodynamics which are the magnetic field $\vec{B}$ the velocity field $\vec{v}$ and the density ρ. Kats [8] has generalized Moffatt's work for compressible non barotropic indexcompressible flows but without reducing the number of functions and the computational load. Moreover, Kats has shown in a series of remarkable papers that the variables he suggested can be utilized to describe the motion of arbitrary discontinuity surfaces [9, 10, 11]. Sakurai [12] has introduced a two function Eulerian variational principle for force-free magnetohydrodynamics based on "Euler potentials" and used it as a basis of a numerical scheme, his method is discussed in a book by Sturrock [3]. A method of solving the equations for those two variables was introduced by Yang, Sturrock & Antiochos [13]. Yahalom & Lynden-Bell [14, 15] have combined the Lagrangian of Sturrock [3] with the Lagrangian of Sakurai [12] to obtain an Eulerian variational principle depending on only six functions. The vanishing of the variational derivatives of this Lagrangian entail all the equations needed to describe barotropic magnetohydrodynamics without any additional constraints. The equations obtained resemble the Hamiltonian equations of Frenkel, Levich & Stilman [16] (see also [17]), the same Hamiltonian equations were obtained at around the same time independently by Morrison [18] who was concerned about obtaining proper Poisson brackets for magnetohydrodynamics. Furthermore, it was shown by Yahalom & Lynden-Bell [14] that for stationary flows three functions will suffice in order to describe a Lagrangian principle for barotropic magnetohydrodynamics.

The non-singlevaluedness of the functions appearing in the reduced representation of barotropic magnetohydrodynamics was discussed in particular with connection to the topological invariants of magnetic and cross helicities. It was shown that flows with non trivial topologies which have non zero magnetic or cross helicities can be adequately described by the functions of the reduced representation provided that some of them are non-single valued [14, 15]. The cross helicity per unit flux was shown to be equal to the discontinuity of the function ν, this discontinuity was shown to be a conserved quantity along the flow. The magnetic helicity per unit flux was shown to be equal to the discontinuity of another function ζ. It should be mentioned that the existence of non single valued functions in the description of toroidal magnetohydrodynamics was first suggested by [19].

In the current chapter we will also discuss the results of [20] which improve upon the results of Yahalom & Lynden-Bell [14, 15] and show that four functions are enough to describe a general non stationary barotropic magnetohydrodynamics in accordance with [20].

This chapter is concerned only with barotropic magnetohydrodynamics. Variational principles of non barotropic magnetohydrodynamics can be found in the work of Bekenstein & Oron [21] in terms of 15 functions and V.A. Kats [8] in terms of 20 functions. The author of this chapter suspect that this number can be somewhat reduced. Moreover, A. V. Kats in a remarkable paper [22] (section IV,E) has shown that there is a large symmetry group (gauge freedom) associated with the choice of those functions, this implies that the number of degrees of freedom can be reduced.

Anticipated applications of this study both to linear and non-linear stability analysis of known barotropic magnetohydrodynamic configurations [23, 24] and for designing efficient numerical schemes for integrating the equations of magnetohydrodynamics [13, 35, 36, 37].

The plan of this chapter is as follows: first we introduce the standard notations and equations of barotropic magnetohydrodynamics . Next we review the Lagrangian variational principle of barotropic magnetohydrodynamics. This is followed by a review of the Eulerian variational principles of force-free magnetohydrodynamics. After those introductory subsections we will present the four function Eulerian variational principles for non-stationary magnetohydrodynamics. A derivation of the canonical momenta of the generalized coordinates appearing in the Lagrangian allows us to derive the system's Hamiltonian which resembles the Hamiltonian introduced by Frenkel, Levich & Stilman [16]. This is followed by the derivation of a variational principle for stationary magnetohydrodynamics. The discussion related to the magnetohydrodynamic topological constants concludes our chapter.

5.2 The standard formulation of barotropic magnetohydrodynamics

5.2.1 Basic equations

The standard set of equations solved for barotropic magnetohydrodynamics are given below:

$$\frac{\partial \vec{B}}{\partial t} = \vec{\nabla} \times (\vec{v} \times \vec{B}), \tag{5.1}$$

$$\vec{\nabla} \cdot \vec{B} = 0, \tag{5.2}$$

$$\frac{\partial \rho}{\partial t} + \vec{\nabla} \cdot (\rho \vec{v}) = 0, \tag{5.3}$$

$$\rho \frac{d\vec{v}}{dt} = \rho(\frac{\partial \vec{v}}{\partial t} + (\vec{v} \cdot \vec{\nabla})\vec{v}) = -\vec{\nabla} p(\rho) + \frac{(\vec{\nabla} \times \vec{B}) \times \vec{B}}{4\pi}. \tag{5.4}$$

The following notations are utilized: $\vec{B}$ is the magnetic field vector, $\vec{v}$ is the velocity field vector and ρ is the fluid density. Finally $p(\rho)$ is the pressure which we assume depends on the density alone (barotropic case). The justification for those equations and the conditions under which they apply can be found in standard books on magnetohydrodynamics (see for example [3]). Equation (5.1) describes the fact that the magnetic field lines are moving with the fluid elements ("frozen" magnetic field lines) and is analogue to equation (1.3) describing "frozen" vortex lines which are of course no longer "frozen" in the magnetohydrodynamics case. Equation (5.2) describes the fact that the magnetic field is solenoidal, equation (5.3) describes the conservation of mass and equation (5.4) is the Euler equation for a fluid in which both pressure and Lorentz magnetic forces apply. The term:

$$\vec{J} = \frac{\vec{\nabla} \times \vec{B}}{4\pi}, \tag{5.5}$$

is the electric current density which is not connected to any mass flow. The number of independent variables for which one needs to solve is seven ($\vec{v}, \vec{B}, \rho$) and the number of equations (5.1,5.3,5.4) is also seven. Notice that equation

(5.2) is a condition on the initial $\vec{B}$ field and is satisfied automatically for any other time due to equation (5.1). Also notice that $p(\rho)$ is not a variable rather it is a given function of ρ.

5.2.2 Lagrangian variational principle of magnetohydrodynamics

A Lagrangian variational principle for barotropic magnetohydrodynamics has been discussed by a number of authors (see for example [3]) and an outline of this approach is given below. Consider the action:

$$\begin{aligned} A &\equiv \int \mathcal{L} d^3x dt, \\ \mathcal{L} &\equiv \rho(\frac{1}{2}\vec{v}^2 - \varepsilon(\rho)) - \frac{\vec{B}^2}{8\pi}, \end{aligned} \tag{5.6}$$

in which $\varepsilon(\rho)$ is the specific internal energy. A variation in any quantity F for a fixed position $\vec{r}$ is denoted as δF hence:

$$\begin{aligned} \delta A &= \int \delta\mathcal{L} d^3x dt, \\ \delta\mathcal{L} &= \delta\rho(\frac{1}{2}\vec{v}^2 - w(\rho)) + \rho\vec{v}\cdot\delta\vec{v} - \frac{\vec{B}\cdot\delta\vec{B}}{4\pi}, \end{aligned} \tag{5.7}$$

in which $w = \frac{\partial(\varepsilon\rho)}{\partial\rho}$ is the specific enthalpy.

A change in a position of a fluid element located at a position $\vec{r}$ at time t is given by $\vec{\xi}(\vec{r},t)$. A mass conserving variation of ρ takes the form:

$$\delta\rho = -\vec{\nabla}\cdot(\rho\vec{\xi}) \tag{5.8}$$

and a magnetic flux conserving variation takes the form:

$$\delta\vec{B} = \vec{\nabla}\times(\vec{\xi}\times\vec{B}). \tag{5.9}$$

A change involving both a local variation coupled with a change of element position of the quantity F is given by:

$$\Delta F = \delta F + (\vec{\xi}\cdot\vec{\nabla})F, \tag{5.10}$$

hence

$$\Delta\vec{v} = \delta\vec{v} + (\vec{\xi}\cdot\vec{\nabla})\vec{v}. \tag{5.11}$$

However, since:

$$\Delta\vec{v} = \Delta\frac{d\vec{r}}{dt} = \frac{d\Delta\vec{r}}{dt} = \frac{d\vec{\xi}}{dt}. \tag{5.12}$$

We obtain:

$$\delta\vec{v} = \frac{d\vec{\xi}}{dt} - (\vec{\xi}\cdot\vec{\nabla})\vec{v} = \frac{\partial\vec{\xi}}{\partial t} + (\vec{v}\cdot\vec{\nabla})\vec{\xi} - (\vec{\xi}\cdot\vec{\nabla})\vec{v}. \tag{5.13}$$

Introducing the result of equations (5.8,5.9,5.13) into equation (5.7) and integrating by parts we arrive at the result:

$$\begin{aligned}\delta A &= \int d^3x\rho\vec{v}\cdot\vec{\xi}|_{t_0}^{t_1}\\ &+ \int dt\{\oint d\vec{S}\cdot[-\rho\vec{\xi}(\frac{1}{2}\vec{v}^2 - w(\rho)) + \rho\vec{v}(\vec{v}\cdot\vec{\xi}) + \frac{1}{4\pi}\vec{B}\times(\vec{\xi}\times\vec{B})]\\ &+ \int d^3x\vec{\xi}\cdot[-\rho\vec{\nabla}w - \frac{\partial(\rho\vec{v})}{\partial t} - \frac{\partial(\rho\vec{v}v_k)}{\partial x_k} - \frac{1}{4\pi}\vec{B}\times(\vec{\nabla}\times\vec{B})]\},\end{aligned} \tag{5.14}$$

in which a summation convention is assumed. Taking into account the continuity equation (5.3) we obtain:

$$\begin{aligned}\delta A &= \int d^3x\rho\vec{v}\cdot\vec{\xi}|_{t_0}^{t_1}\\ &+ \int dt\{\oint d\vec{S}\cdot[-\rho\vec{\xi}(\frac{1}{2}\vec{v}^2 - w(\rho)) + \rho\vec{v}(\vec{v}\cdot\vec{\xi}) + \frac{1}{4\pi}\vec{B}\times(\vec{\xi}\times\vec{B})]\\ &+ \int d^3x\vec{\xi}\cdot[-\rho\vec{\nabla}w - \rho\frac{\partial\vec{v}}{\partial t} - \rho(\vec{v}\cdot\vec{\nabla})\vec{v} - \frac{1}{4\pi}\vec{B}\times(\vec{\nabla}\times\vec{B})]\},\end{aligned} \tag{5.15}$$

hence we see that if $\delta A = 0$ for a $\vec{\xi}$ vanishing at the initial and final times and on the surface of the domain but otherwise arbitrary then Euler's equation (5.4) is satisfied (taking into account that in the barotropic case $\vec{\nabla}w = \frac{\vec{\nabla}p}{\rho}$). The reader should notice that the vanishing of $\vec{\xi}$ on the surface of the domain is only a sufficient condition for Euler's equations to be satisfied. It is certainly not a necessary condition. In scenarios such that the boundary is impermeable and perfectly conducting, it suffices to assume that only the normal component of $\vec{\xi}$ is zero at the boundary.

Although the variational principle does give us the correct dynamical equation for an arbitrary $\vec{\xi}$, it has the following deficiencies:

1. Although $\vec{\xi}$ is quite arbitrary the variations of $\delta\rho$ and $\delta\vec{B}$ are not. They are defined by the conditions given in equation (5.8) and equation (5.9). This property is not useful for numerical schemes since $\vec{\xi}$ must be a small quantity.

2. Only equation (5.4) is derived from the variational principle the other equations that are needed: equation (5.1), equation (5.2) and equation (5.3) are separate assumptions. Moreover equation (5.3) is needed in order to derive Euler's equation (5.4) from the variational principle. All this makes the variational principle less useful.

What is desired is a variational principle from which all equations of motion can be derived and for which no assumptions on the variations are needed this will be discussed in the following sections.

5.3 Sakurai's variational principle of force-free magnetohydrodynamics

Force-free magnetohydrodynamics is concerned with the case that both the pressure and inertial terms in Euler equations (5.4) are physically insignificant. Hence the Euler equations can be written in the form:

$$\frac{(\vec{\nabla} \times \vec{B}) \times \vec{B}}{4\pi} = \vec{J} \times \vec{B} = 0. \tag{5.16}$$

In order to describe force-free fields Sakurai [12] has proposed to represent the magnetic field in the following form.

$$\vec{B} = \vec{\nabla}\chi \times \vec{\nabla}\eta. \tag{5.17}$$

Hence $\vec{B}$ is orthogonal both to $\vec{\nabla}\chi$ and $\vec{\nabla}\eta$. A similar representation was suggested by Dungey [25, p. 31] but not in the context of variational analysis. Frenkel, Levich & Stilman [16] has discussed the validity of the above representation and have concluded that for a vector field in the Euclidean space $\mathcal{R}^3$ the above presentation does always exist locally but not always globally. Also Notice that either χ or η (or both) can be **non single valued functions** (see [16] equation 20).

Both χ and η are Clebsch type comoving scalar fields satisfying the equations:

$$\frac{d\chi}{dt} = 0, \qquad \frac{d\eta}{dt} = 0. \tag{5.18}$$

It can be easily shown that provided that $\vec{B}$ is in the form given in equation (5.17), and equation (5.18) is satisfied, then both equation (5.1) and equation (5.2) are satisfied. Since according to equation (5.16) both $\vec{\nabla} \times \vec{B}$ and $\vec{B}$ are parallel it follows that equation (5.16) can be written as:

$$\vec{J} \cdot \vec{\nabla}\chi = 0, \qquad \vec{J} \cdot \vec{\nabla}\eta = 0. \tag{5.19}$$

Sakurai [12] has introduced an action principle from which equation (5.19) can be derived:

$$\begin{aligned} A_S &\equiv \int \mathcal{L}_S d^3x dt, \\ \mathcal{L}_S &\equiv \frac{\vec{B}^2}{8\pi} = \frac{(\vec{\nabla}\chi \times \vec{\nabla}\eta)^2}{8\pi}. \end{aligned} \tag{5.20}$$

Taking the variation of equation (5.20) we obtain:

$$\begin{aligned} \delta A_S &= \int \delta \mathcal{L}_S d^3x dt, \\ \delta \mathcal{L}_S &= \frac{\vec{B}}{4\pi} \cdot (\vec{\nabla}\delta\chi \times \vec{\nabla}\eta + \vec{\nabla}\chi \times \vec{\nabla}\delta\eta). \end{aligned} \tag{5.21}$$

Integrating by parts and using the theorem of Gauss one obtains the result:

$$\begin{aligned} \delta A_S &= \oint d\vec{S} \cdot [(\delta\chi\vec{\nabla}\eta - \delta\eta\vec{\nabla}\chi) \times \frac{\vec{B}}{4\pi}] + \int d\vec{\Sigma} \cdot [([\delta\chi]\vec{\nabla}\eta - [\delta\eta]\vec{\nabla}\chi) \times \frac{\vec{B}}{4\pi}] \\ &+ \int d^3x[\delta\chi(\vec{\nabla}\eta \cdot \vec{J}) - \delta\eta(\vec{\nabla}\chi \cdot \vec{J})] \end{aligned} \tag{5.22}$$

in which $\int d\vec{\Sigma}$ represents an integral along the cut and $[\delta f]$ represents the discontinuity of the variations of non single valued functions. The reason we introduce cuts in the domain is because we are not sure at this stage whether the χ and η functions are single valued or multiple valued. We shall show later that χ can be defined as a single valued function while η can be either single valued or non single valued. In the later case a sufficient condition for the "cut" term to vanish is the usage of a single valued $\delta\eta$, that is we may vary η only using single valued variations. Hence if $\delta A_S = 0$ for arbitrary variation $\delta\chi, \delta\eta$ that vanish on the boundary of the domain (including the cut) one recovers the force-free Euler equations (5.19).

Although this approach is better than the one described in equation (5.6) in the previous section in the sense that the form of the variations $\delta\chi, \delta\eta$ is not constrained, it has some limitations as follows:

1. Sakurai's approach by design is only meant to deal with force-free magnetohydrodynamics; for more general magnetohydrodynamics it is not adequate.

2. Sakurai's action given by equation (5.20) contains all the relevant physical equations only if the configuration is static ($\vec{v} = 0$). If the configuration is not static one needs to supply an additional two equations (5.18) to the variational principle.

5.4 Simplified variational principle of non-stationary barotropic magnetohydrodynamics

In the following section we will combine the approaches described in the previous sections in order to obtain a variational principle of non-stationary barotropic magnetohydrodynamics such that all the relevant barotropic magnetohydrodynamic equations can be derived from using unconstrained variations. The approach is based on a method first introduced by Davydov and Seliger & Whitham [4, 6]. Consider the action:

$$
\begin{aligned}
A &\equiv \int \mathcal{L} d^3x dt, \\
\mathcal{L} &\equiv \mathcal{L}_1 + \mathcal{L}_2, \\
\mathcal{L}_1 &\equiv \rho(\frac{1}{2}\vec{v}^2 - \varepsilon(\rho)) + \frac{\vec{B}^2}{8\pi}, \\
\mathcal{L}_2 &\equiv \nu[\frac{\partial\rho}{\partial t} + \vec{\nabla}\cdot(\rho\vec{v})] - \rho\alpha\frac{d\chi}{dt} - \rho\beta\frac{d\eta}{dt} - \frac{\vec{B}}{4\pi}\cdot(\vec{\nabla}\chi \times \vec{\nabla}\eta). \quad (5.23)
\end{aligned}
$$

Obviously ν, α, β (the α, β defined in this section are not related to the variables with the same name defined in the previous section) are Lagrange multipliers which were inserted in such a way that the variational principle will yield the following equations:

$$\frac{\partial \rho}{\partial t} + \vec{\nabla} \cdot (\rho \vec{v}) = 0,$$
$$\rho \frac{d\chi}{dt} = 0,$$
$$\rho \frac{d\eta}{dt} = 0. \qquad (5.24)$$

It **is not** assumed that ν, α, β are single valued. Provided ρ is not null those are just the continuity equation (5.3) and the conditions that Sakurai's functions are comoving as in equation (5.18). Taking the variational derivative with respect to $\vec{B}$ we see that

$$\vec{B} = \hat{\vec{B}} \equiv \vec{\nabla}\chi \times \vec{\nabla}\eta. \qquad (5.25)$$

Hence $\vec{B}$ is in Sakurai's form and satisfies equation (5.2). By virtue of equations (5.24) we see that $\vec{B}$ must also satisfy equation (5.1). For the time being we have showed that all the equations of barotropic magnetohydrodynamics can be obtained from the above variational principle except Euler's equations. We will now show that Euler's equations can be derived from the above variational principle as well. Let us take an arbitrary variational derivative of the above action with respect to $\vec{v}$, this will result in:

$$\delta_{\vec{v}} A = \int d^3x dt \rho \delta\vec{v} \cdot [\vec{v} - \vec{\nabla}\nu - \alpha\vec{\nabla}\chi - \beta\vec{\nabla}\eta] + \oint d\vec{S} \cdot \delta\vec{v} \rho \nu + \int d\vec{\Sigma} \cdot \delta\vec{v} \rho [\nu]. \qquad (5.26)$$

The integral $\oint d\vec{S} \cdot \delta\vec{v} \rho \nu$ vanishes in many physical scenarios. In the case of astrophysical flows this integral will vanish since $\rho = 0$ on the flow boundary, in the case of a fluid contained in a vessel no flux boundary conditions $\delta\vec{v} \cdot \hat{n} = 0$ are induced ($\hat{n}$ is a unit vector normal to the boundary). The surface integral $\int d\vec{\Sigma}$ on the cut of ν vanishes in the case that the flow has zero cross helicity (see section 5.7) since in this case ν is single valued and $[\nu] = 0$. In the case that that the flow has non zero cross helicity, ν is not single valued (see section 5.7), in this case only a Kutta type velocity perturbation [15] in which the velocity perturbation is parallel to the cut will cause the cut integral to vanish.

Provided that the surface integrals do vanish and that $\delta_{\vec{v}} A = 0$ for an arbitrary velocity perturbation we see that $\vec{v}$ must have the following form:

$$\vec{v} = \hat{\vec{v}} \equiv \vec{\nabla}\nu + \alpha\vec{\nabla}\chi + \beta\vec{\nabla}\eta. \tag{5.27}$$

Let us now take the variational derivative with respect to the density ρ we obtain:

$$\begin{aligned} \delta_\rho A &= \int d^3x dt \delta\rho[\frac{1}{2}\vec{v}^2 - w - \frac{\partial\nu}{\partial t} - \vec{v}\cdot\vec{\nabla}\nu] \\ &+ \oint d\vec{S}\cdot\vec{v}\delta\rho\nu + \int d\vec{\Sigma}\cdot\vec{v}\delta\rho[\nu] + \int d^3x\nu\delta\rho|_{t_0}^{t_1}. \end{aligned} \tag{5.28}$$

Hence provided that $\oint d\vec{S}\cdot\vec{v}\delta\rho\nu$ vanishes on the boundary of the domain and $\int d\vec{\Sigma}\cdot\vec{v}\delta\rho[\nu]$ vanishes on the cut of ν in the case that ν is not single valued[1] and in initial and final times the following equation must be satisfied:

$$\frac{d\nu}{dt} = \frac{1}{2}\vec{v}^2 - w. \tag{5.29}$$

Finally we have to calculate the variation with respect to both χ and η this will lead us to the following results:

$$\begin{aligned} \delta_\chi A &= \int d^3x dt \delta\chi[\frac{\partial(\rho\alpha)}{\partial t} + \vec{\nabla}\cdot(\rho\alpha\vec{v}) - \vec{\nabla}\eta\cdot\vec{J}] \\ &+ \oint d\vec{S}\cdot[\frac{\vec{B}}{4\pi}\times\vec{\nabla}\eta - \vec{v}\rho\alpha]\delta\chi \\ &+ \int d\vec{\Sigma}\cdot[\frac{\vec{B}}{4\pi}\times\vec{\nabla}\eta - \vec{v}\rho\alpha][\delta\chi] - \int d^3x\rho\alpha\delta\chi|_{t_0}^{t_1}, \end{aligned} \tag{5.30}$$

$$\begin{aligned} \delta_\eta A &= \int d^3x dt \delta\eta[\frac{\partial(\rho\beta)}{\partial t} + \vec{\nabla}\cdot(\rho\beta\vec{v}) + \vec{\nabla}\chi\cdot\vec{J}] \\ &+ \oint d\vec{S}\cdot[\vec{\nabla}\chi\times\frac{\vec{B}}{4\pi} - \vec{v}\rho\beta]\delta\eta \\ &+ \int d\vec{\Sigma}\cdot[\vec{\nabla}\chi\times\frac{\vec{B}}{4\pi} - \vec{v}\rho\beta][\delta\eta] - \int d^3x\rho\beta\delta\eta|_{t_0}^{t_1}. \end{aligned} \tag{5.31}$$

[1]Which entails either a Kutta type condition for the velocity or a vanishing density perturbation on the cut.

Provided that the correct temporal and boundary conditions are met with respect to thc variations $\delta\chi$ and $\delta\eta$ on the domain boundary and on the cuts in the case that some (or all) of the relevant functions are non single valued. we obtain the following set of equations:

$$\frac{d\alpha}{dt} = \frac{\vec{\nabla}\eta \cdot \vec{J}}{\rho}, \qquad \frac{d\beta}{dt} = -\frac{\vec{\nabla}\chi \cdot \vec{J}}{\rho}, \tag{5.32}$$

in which the continuity equation (5.3) was taken into account. By correct temporal conditions we mean that both $\delta\eta$ and $\delta\chi$ vanish at initial and final times. As for boundary conditions which are sufficient to make the boundary term vanish we can consider the case that the boundary is at infinity and both $\vec{B}$ and ρ vanish. Another possibility is that the boundary is impermeable and perfectly conducting. A sufficient condition for the integral over the "cuts" to vanish is to use variations $\delta\eta$ and $\delta\chi$ which are single valued. It will be shown later that χ can always be taken to be single valued, hence taking $\delta\chi$ to be single valued is no restriction at all. In some topologies η is not single valued and in those cases a single valued restriction on $\delta\eta$ is sufficient to make the cut term null.

5.4.1 Euler's equations

We shall now show that a velocity field given by equation (5.27), such that the equations for $\alpha, \beta, \chi, \eta, \nu$ satisfy the corresponding equations (5.24,5.29,5.32) must satisfy Euler's equations. Let us calculate the material derivative of $\vec{v}$:

$$\frac{d\vec{v}}{dt} = \frac{d\vec{\nabla}\nu}{dt} + \frac{d\alpha}{dt}\vec{\nabla}\chi + \alpha\frac{d\vec{\nabla}\chi}{dt} + \frac{d\beta}{dt}\vec{\nabla}\eta + \beta\frac{d\vec{\nabla}\eta}{dt}. \tag{5.33}$$

It can be easily shown that:

$$\begin{aligned}
\frac{d\vec{\nabla}\nu}{dt} &= \vec{\nabla}\frac{d\nu}{dt} - \vec{\nabla}v_k\frac{\partial\nu}{\partial x_k} = \vec{\nabla}(\frac{1}{2}\vec{v}^2 - w) - \vec{\nabla}v_k\frac{\partial\nu}{\partial x_k},\\
\frac{d\vec{\nabla}\eta}{dt} &= \vec{\nabla}\frac{d\eta}{dt} - \vec{\nabla}v_k\frac{\partial\eta}{\partial x_k} = -\vec{\nabla}v_k\frac{\partial\eta}{\partial x_k},\\
\frac{d\vec{\nabla}\chi}{dt} &= \vec{\nabla}\frac{d\chi}{dt} - \vec{\nabla}v_k\frac{\partial\chi}{\partial x_k} = -\vec{\nabla}v_k\frac{\partial\chi}{\partial x_k}.
\end{aligned} \tag{5.34}$$

In which x_k is a Cartesian coordinate and a summation convention is assumed. Equations (5.24,5.29) where used in the above derivation. Inserting the result from equations (5.34,5.32) into equation (5.33) yields:

$$\begin{aligned}\frac{d\vec{v}}{dt} &= -\vec{\nabla}v_k(\frac{\partial\nu}{\partial x_k}+\alpha\frac{\partial\chi}{\partial x_k}+\beta\frac{\partial\eta}{\partial x_k})+\vec{\nabla}(\frac{1}{2}\vec{v}^2-w)\\ &+ \frac{1}{\rho}((\vec{\nabla}\eta\cdot\vec{J})\vec{\nabla}\chi-(\vec{\nabla}\chi\cdot\vec{J})\vec{\nabla}\eta)\\ &= -\vec{\nabla}v_k v_k+\vec{\nabla}(\frac{1}{2}\vec{v}^2-w)+\frac{1}{\rho}\vec{J}\times(\vec{\nabla}\chi\times\vec{\nabla}\eta)\\ &= -\frac{\vec{\nabla}p}{\rho}+\frac{1}{\rho}\vec{J}\times\vec{B}.\end{aligned} \tag{5.35}$$

In which we have used both equation (5.27) and equation (5.25) in the above derivation. This of course proves that the barotropic Euler equations can be derived from the action given in equation (5.23) and hence all the equations of barotropic magnetohydrodynamics can be derived from the above action without restricting the variations in any way except on the relevant boundaries and cuts. The reader should take into account that the topology of the magnetohydrodynamic flow is conserved, hence cuts must be introduced into the calculation as initial conditions.

5.4.2 Simplified action

The reader of this chapter might argue here that the chapter is misleading. The author has declared that he is going to present a simplified action for barotropic magnetohydrodynamics instead he has added five more functions $\alpha, \beta, \chi, \eta, \nu$ to the standard set $\vec{B}, \vec{v}, \rho$. In the following I will show that this is not so and the action given in equation (5.23) in a form suitable for a pedagogic presentation can indeed be simplified. It is easy to show that the Lagrangian density appearing in equation (5.23) can be written in the form:

$$\begin{aligned}\mathcal{L} &= -\rho[\frac{\partial\nu}{\partial t}+\alpha\frac{\partial\chi}{\partial t}+\beta\frac{\partial\eta}{\partial t}+\varepsilon(\rho)]+\frac{1}{2}\rho[(\vec{v}-\hat{\vec{v}})^2-(\hat{\vec{v}})^2]\\ &+ \frac{1}{8\pi}[(\vec{B}-\hat{\vec{B}})^2-(\hat{\vec{B}})^2]+\frac{\partial(\nu\rho)}{\partial t}+\vec{\nabla}\cdot(\nu\rho\vec{v}).\end{aligned} \tag{5.36}$$

In which $\hat{\vec{v}}$ is a shorthand notation for $\vec{\nabla}\nu+\alpha\vec{\nabla}\chi+\beta\vec{\nabla}\eta$ (see equation (5.27)) and $\hat{\vec{B}}$ is a shorthand notation for $\vec{\nabla}\chi\times\vec{\nabla}\eta$ (see equation (5.25)). Thus $\mathcal{L}$

has four contributions:

$$\begin{aligned}
\mathcal{L} &= \hat{\mathcal{L}} + \mathcal{L}_{\vec{v}} + \mathcal{L}_{\vec{B}} + \mathcal{L}_{boundary}, \\
\hat{\mathcal{L}} &\equiv -\rho\left[\frac{\partial \nu}{\partial t} + \alpha\frac{\partial \chi}{\partial t} + \beta\frac{\partial \eta}{\partial t} + \varepsilon(\rho) + \frac{1}{2}(\vec{\nabla}\nu + \alpha\vec{\nabla}\chi + \beta\vec{\nabla}\eta)^2\right] \\
&- \frac{1}{8\pi}(\vec{\nabla}\chi \times \vec{\nabla}\eta)^2 \\
\mathcal{L}_{\vec{v}} &\equiv \frac{1}{2}\rho(\vec{v} - \hat{\vec{v}})^2, \\
\mathcal{L}_{\vec{B}} &\equiv \frac{1}{8\pi}(\vec{B} - \hat{\vec{B}})^2, \\
\mathcal{L}_{boundary} &\equiv \frac{\partial(\nu\rho)}{\partial t} + \vec{\nabla}\cdot(\nu\rho\vec{v}). \qquad (5.37)
\end{aligned}$$

The only term containing $\vec{v}$ is[2] $\mathcal{L}_{\vec{v}}$, it can easily be seen that this term will lead, after we nullify the variational derivative with respect to $\vec{v}$, to equation (5.27) but will otherwise have no contribution to other variational derivatives. Similarly the only term containing $\vec{B}$ is $\mathcal{L}_{\vec{B}}$ and it can easily be seen that this term will lead, after we nullify the variational derivative, to equation (5.25) but will have no contribution to other variational derivatives. Also notice that the term $\mathcal{L}_{boundary}$ contains only complete partial derivatives and thus can not contribute to the equations although it can change the boundary conditions. Hence we see that equations (5.24), equation (5.29) and equations (5.32) can be derived using the Lagrangian density $\hat{\mathcal{L}}[\alpha, \beta, \chi, \eta, \nu, \rho]$ in which $\hat{\vec{v}}$ replaces $\vec{v}$ and $\hat{\vec{B}}$ replaces $\vec{B}$ in the relevant equations. Furthermore, after integrating the six equations (5.24,5.29,5.32) we can insert the potentials $\alpha, \beta, \chi, \eta, \nu$ into equations (5.27) and (5.25) to obtain the physical quantities $\vec{v}$ and $\vec{B}$. Hence, the general barotropic magnetohydrodynamic problem is reduced from seven equations (5.1,5.3,5.4) and the additional constraint (5.2) to a problem of six first order (in the temporal derivative) unconstrained equations. Moreover, the entire set of equations can be derived from the Lagrangian density $\hat{\mathcal{L}}$ which is what we were aiming to prove.

It should be emphasized that for non trivial topologies it is necessary to assume that some of the variables introduced in the non stationary formalism are non single-valued.

---

[2] $\mathcal{L}_{boundary}$ also depends on $\vec{v}$ but being a boundary term is space and time it does not contribute to the derived equations

5.4.3 A Simpler Action for Barotropic Magnetohydrodynamics

Can we obtain a further reduction of barotropic magnetohydrodynamics? Can we formulate magnetohydrodynamics with less than the six functions $\alpha, \beta, \chi, \eta, \nu, \rho$? The answer is yes, in fact four functions χ, η, ν, ρ will suffice. To see this we may write the two equations (5.24) as equations for α, β that is:

$$\frac{d\chi}{dt} = \frac{\partial\chi}{\partial t} + \vec{v}\cdot\vec{\nabla}\chi = \frac{\partial\chi}{\partial t} + (\vec{\nabla}\nu + \alpha\vec{\nabla}\chi + \beta\vec{\nabla}\eta)\cdot\vec{\nabla}\chi = 0,$$
$$\frac{d\eta}{dt} = \frac{\partial\eta}{\partial t} + \vec{v}\cdot\vec{\nabla}\eta = \frac{\partial\eta}{\partial t} + (\vec{\nabla}\nu + \alpha\vec{\nabla}\chi + \beta\vec{\nabla}\eta)\cdot\vec{\nabla}\eta = 0, \quad (5.38)$$

in which we have used equation (5.27). Solving for α, β we obtain:

$$\alpha[\chi,\eta,\nu] = \frac{(\vec{\nabla}\eta)^2(\frac{\partial\chi}{\partial t} + \vec{\nabla}\nu\cdot\vec{\nabla}\chi) - (\vec{\nabla}\eta\cdot\vec{\nabla}\chi)(\frac{\partial\eta}{\partial t} + \vec{\nabla}\nu\cdot\vec{\nabla}\eta)}{(\vec{\nabla}\eta\cdot\vec{\nabla}\chi)^2 - (\vec{\nabla}\eta)^2(\vec{\nabla}\chi)^2}$$
$$\beta[\chi,\eta,\nu] = \frac{(\vec{\nabla}\chi)^2(\frac{\partial\eta}{\partial t} + \vec{\nabla}\nu\cdot\vec{\nabla}\eta) - (\vec{\nabla}\eta\cdot\vec{\nabla}\chi)(\frac{\partial\chi}{\partial t} + \vec{\nabla}\nu\cdot\vec{\nabla}\chi)}{(\vec{\nabla}\eta\cdot\vec{\nabla}\chi)^2 - (\vec{\nabla}\eta)^2(\vec{\nabla}\chi)^2}. \quad (5.39)$$

Hence α and β are not free variables any more, but depend on χ, η, ν. Moreover, the velocity $\vec{v}$ now depends on the same three variables χ, η, ν:

$$\vec{v} = \vec{\nabla}\nu + \alpha[\chi,\eta,\nu]\vec{\nabla}\chi + \beta[\chi,\eta,\nu]\vec{\nabla}\eta. \quad (5.40)$$

Since $\vec{v}$ is given now by equation (5.40) it follows that the two equations (5.24) are satisfied identically and need not be derived from a variational principle. The above equation can be somewhat simplified resulting in:

$$\vec{v} = \vec{\nabla}\nu + \frac{1}{\vec{B}^2}[\frac{\partial\eta}{\partial t}\vec{\nabla}\chi - \frac{\partial\chi}{\partial t}\vec{\nabla}\eta + \vec{\nabla}\nu\times\vec{B}]\times\vec{B}$$
$$= \frac{1}{\vec{B}^2}[(\frac{\partial\eta}{\partial t}\vec{\nabla}\chi - \frac{\partial\chi}{\partial t}\vec{\nabla}\eta)\times\vec{B} + \vec{B}(\vec{\nabla}\nu\cdot\vec{B})] \equiv \vec{v}_\perp + \vec{v}_\parallel \quad (5.41)$$

Hence the velocity $\vec{v}$ is partitioned naturally into two components one which is parallel to the magnetic field and another one which is perpendicular to it. Crossing the velocity representation (5.40) with the appropriate gradient will lead to the results:

$$\alpha = \frac{\vec{\nabla}\eta \cdot (\vec{B} \times (\vec{v} - \vec{\nabla}\nu))}{\vec{B}^2}$$
$$\beta = -\frac{\vec{\nabla}\chi \cdot (\vec{B} \times (\vec{v} - \vec{\nabla}\nu))}{\vec{B}^2}. \quad (5.42)$$

The reader should notice that the above quantities become singular for $\vec{B} = 0$, hence the formalism is only adequate for describing flows for which $\vec{B} \neq 0$. If the magnetic field vanishes at infinity this is not an obstacle from the present formalism point of view since the velocity field need not be defined at infinity where there is no flow. For flow without magnetic fields the present formalism is not appropriate and other variational economic formalisms may be suggested. It is necessary to mention that to describe Eulerian fluids with the help of Clebsch potentials one needs three functions which are nothing more than initial positions of fluid particles [26]. Finally equations (5.39) should be substituted into equation (5.23) to obtain a Lagrangian density $\mathcal{L}$ in terms of χ, η, ν, ρ:

$$\mathcal{L}[\chi, \eta, \nu, \rho] = \rho[\frac{1}{2}\vec{v}^2 - \frac{d\nu}{dt} - \varepsilon(\rho)] - \frac{\vec{B}^2}{8\pi} \quad (5.43)$$

where $\vec{v}$ is given by equation (5.41) and $\vec{B}$ by equation (5.17). Or more explicitly as:

$$\mathcal{L}[\chi, \eta, \nu, \rho] = \frac{1}{2}\frac{\rho}{(\vec{\nabla}\chi \times \vec{\nabla}\eta)^2}[\vec{\nabla}\eta\frac{\partial \chi}{\partial t} - \vec{\nabla}\chi\frac{\partial \eta}{\partial t} + (\vec{\nabla}\chi \times \vec{\nabla}\eta) \times \vec{\nabla}\nu]^2$$
$$- \rho[\frac{\partial \nu}{\partial t} + \frac{1}{2}(\vec{\nabla}\nu)^2 + \varepsilon(\rho)] - \frac{(\vec{\nabla}\chi \times \vec{\nabla}\eta)^2}{8\pi}. \quad (5.44)$$

This Lagrangian density admits an infinite symmetry group of transformations of the form:

$$\hat{\eta} = \hat{\eta}(\chi, \eta), \qquad \hat{\chi} = \hat{\chi}(\chi, \eta), \quad (5.45)$$

provided that the absolute value of the Jacobian of these transformation is unity:

$$\left|\frac{\partial(\hat{\eta}, \hat{\chi})}{\partial(\eta, \chi)}\right| = 1. \quad (5.46)$$

In particular the Lagrangian density admits an exchange symmetry:

$$\hat{\eta} = \chi, \qquad \hat{\chi} = \eta. \quad (5.47)$$

Taking the variational derivatives to zero for arbitrary variations of equation (5.44) leads to the following set of equations:

$$\frac{\partial \rho}{\partial t} + \vec{\nabla} \cdot (\rho \vec{v}) = 0, \tag{5.48}$$

$$\frac{d\nu}{dt} = \frac{1}{2}\vec{v}^2 - w, \tag{5.49}$$

$$\frac{d\alpha[\chi,\eta,\nu]}{dt} = \frac{\vec{\nabla}\eta \cdot \vec{J}}{\rho}, \tag{5.50}$$

$$\frac{d\beta[\chi,\eta,\nu]}{dt} = -\frac{\vec{\nabla}\chi \cdot \vec{J}}{\rho}. \tag{5.51}$$

Those equations should be solved for χ, η, ν, ρ

Equations (5.50,5.51) contain a complicated linear combination of the second derivatives $\frac{\partial^2 \chi}{\partial t^2}$ and $\frac{\partial^2 \eta}{\partial t^2}$. This is inconvenient numerically therefore the following approach is recommended. Taking the partial temporal derivatives of the two χ and η equations appearing in the equation set (5.24) (after dividing by ρ) we obtain:

$$\frac{\partial^2 \chi}{\partial t^2} + \frac{\partial \vec{v}}{\partial t} \cdot \vec{\nabla}\chi + \vec{v} \cdot \vec{\nabla}\frac{\partial \chi}{\partial t} = 0, \qquad \frac{\partial^2 \eta}{\partial t^2} + \frac{\partial \vec{v}}{\partial t} \cdot \vec{\nabla}\eta + \vec{v} \cdot \vec{\nabla}\frac{\partial \eta}{\partial t} = 0. \tag{5.52}$$

Using the expression $\frac{\partial \vec{v}}{\partial t}$ from equation (5.35) we obtain an explicit expression for the second derivatives of the form:

$$\begin{aligned} \frac{\partial^2 \chi}{\partial t^2} &= ((\vec{v} \cdot \vec{\nabla})\vec{v} + \vec{\nabla} w - \frac{1}{\rho}\vec{J} \times \vec{B}) \cdot \vec{\nabla}\chi - \vec{v} \cdot \vec{\nabla}\frac{\partial \chi}{\partial t} \\ \frac{\partial^2 \eta}{\partial t^2} &= ((\vec{v} \cdot \vec{\nabla})\vec{v} + \vec{\nabla} w - \frac{1}{\rho}\vec{J} \times \vec{B}) \cdot \vec{\nabla}\eta - \vec{v} \cdot \vec{\nabla}\frac{\partial \eta}{\partial t}. \end{aligned} \tag{5.53}$$

Hence we arrived at a four function formalism for barotropic magnetohydrodynamics which can be derived from a Lagrangian. The equations to be solved for those functions are equations (5.48, 5.49) and equations (5.53). Notice, however, that this formalism contains two first order equations and two second order equations, while our previous six function formalism [14] contained six first order equations.

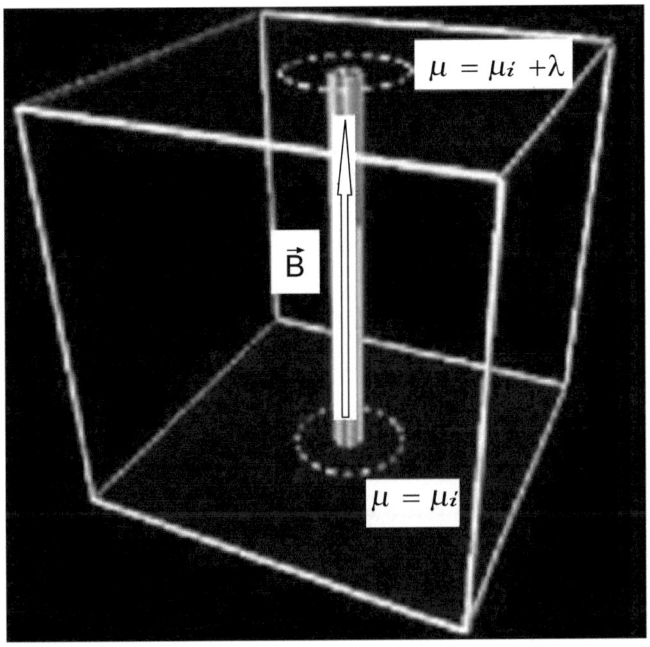

Figure 5.1: A thin tube surrounding a magnetic field line

5.4.4 The inverse problem

In the previous subsection we have shown that given a set of functions $\alpha, \beta, \chi, \eta, \nu$ (or a reduced set χ, η, ν) satisfying the set of equations described in the previous subsections, one can insert those functions into equation (5.27) and equation (5.25) to obtain the physical velocity $\vec{v}$ and magnetic field $\vec{B}$. In this subsection we will address the inverse problem that is, suppose we are given the quantities $\vec{v}, \vec{B}$ and ρ how can one calculate the potentials $\alpha, \beta, \chi, \eta, \nu$? The treatment in this section will follow closely an analogous treatment for non-magnetic fluid dynamics given by Lynden-Bell & Katz [7].

Consider a thin tube surrounding a magnetic field line as described in figure 5.1, the magnetic flux contained within the tube is:

$$\Delta\Phi = \int \vec{B} \cdot d\vec{S} \tag{5.54}$$

and the mass contained with the tube is:

$$\Delta M = \int \rho d\vec{l} \cdot d\vec{S}, \tag{5.55}$$

in which dl is a length element along the tube. Since the magnetic field lines move with the flow by virtue of equation (5.1) both the quantities $\Delta\Phi$ and

ΔM are conserved and since the tube is thin we may define the conserved magnetic load:

$$\lambda = \frac{\Delta M}{\Delta \Phi} = \oint \frac{\rho}{B} dl, \tag{5.56}$$

in which the above integral is performed along the field line. Obviously the parts of the line which go out of the flow to regions in which $\rho = 0$ have a null contribution to the integral. Notice that λ is a **single valued** function that can be measured in principle. Since λ is conserved it satisfies the equation:

$$\frac{d\lambda}{dt} = 0. \tag{5.57}$$

By construction surfaces of constant magnetic load move with the flow and contain magnetic field lines. Hence the gradient to such surfaces must be orthogonal to the field line:

$$\vec{\nabla}\lambda \cdot \vec{B} = 0. \tag{5.58}$$

For a discussion of the possibility of surface and volume filling fields and the complications in the definition of λ then see subsection 5.6.3. Now consider an arbitrary comoving point on the magnetic field line and denote it by i, and consider an additional comoving point on the magnetic field line and denote it by r. The integral:

$$\mu(r) = \int_i^r \frac{\rho}{B} dl + \mu(i), \tag{5.59}$$

is also a conserved quantity which we may denote following Lynden-Bell & Katz [7] as the magnetic metage. $\mu(i)$ is an arbitrary number which can be chosen differently for each magnetic line. By construction:

$$\frac{d\mu}{dt} = 0. \tag{5.60}$$

Also it is easy to see that by differentiating along the magnetic field line we obtain:

$$\vec{\nabla}\mu \cdot \vec{B} = \rho. \tag{5.61}$$

Notice that μ will be generally a **non single valued** function, we will show later in this chapter that symmetry to translations in μ will generate through the Noether theorem the conservation of the magnetic cross helicity.

At this point we have two comoving coordinates of flow, namely λ, μ obviously in a three dimensional flow we also have a third coordinate. However,

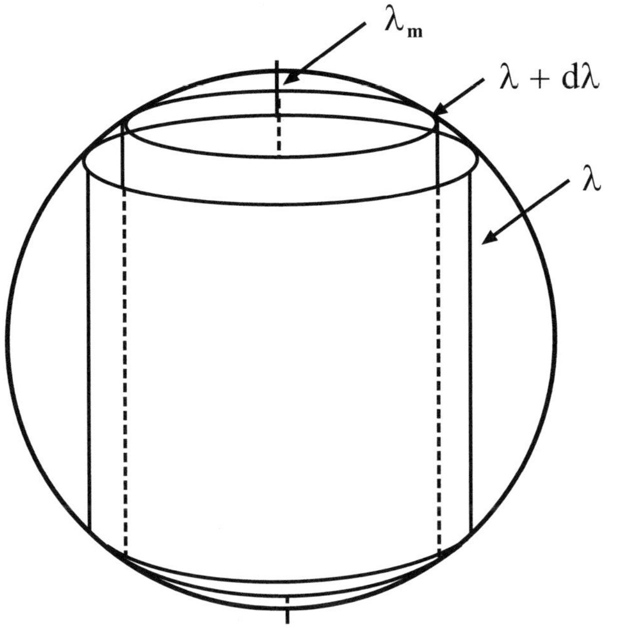

Figure 5.2: Surfaces of constant load

before defining the third coordinate we will find it useful to work not directly with λ but with a function of λ. Now consider the magnetic flux within a surface of constant load $\Phi(\lambda)$ as described in figure 5.2 (the figure was given by Lynden-Bell & Katz [7]). The magnetic flux is a conserved quantity and depends only on the load λ of the surrounding surface. Now we define the quantity:

$$\chi = \frac{\Phi(\lambda)}{2\pi}. \tag{5.62}$$

Obviously χ satisfies the equations:

$$\frac{d\chi}{dt} = 0, \qquad \vec{B} \cdot \vec{\nabla}\chi = 0, \tag{5.63}$$

we will immediately show that this function is identical to Sakurai's function defined in equation (5.17). Let us now define an additional comoving coordinate η^* since $\vec{\nabla}\mu$ is not orthogonal to the $\vec{B}$ lines we can choose $\vec{\nabla}\eta^*$ to be orthogonal to the $\vec{B}$ lines and not be in the direction of the $\vec{\nabla}\chi$ lines, that is we choose η^* not to depend only on χ. Since both $\vec{\nabla}\eta^*$ and $\vec{\nabla}\chi$ are orthogonal to $\vec{B}$, $\vec{B}$ must take the form:

$$\vec{B} = A\vec{\nabla}\chi \times \vec{\nabla}\eta^*. \tag{5.64}$$

However, using equation (5.2) we have:

$$\vec{\nabla} \cdot \vec{B} = \vec{\nabla} A \cdot (\vec{\nabla}\chi \times \vec{\nabla}\eta^*) = 0. \tag{5.65}$$

Which implies that A is a function of χ, η^*. Now we can define a new comoving function η such that:

$$\eta = \int_0^{\eta^*} A(\chi, \eta^{'*}) d\eta^{'*}, \qquad \frac{d\eta}{dt} = 0. \tag{5.66}$$

In terms of this function we recover the Sakurai presentation defined in equation (5.17):

$$\vec{B} = \vec{\nabla}\chi \times \vec{\nabla}\eta. \tag{5.67}$$

Hence we have shown how χ, η can be constructed for a known $\vec{B}, \rho$. Notice however, that η is defined in a non unique way since one can redefine η for example by performing the following transformation: $\eta \rightarrow \eta + f(\chi)$ in which $f(\chi)$ is an arbitrary function. The comoving coordinates χ, η serve as labels of the magnetic field lines. Moreover the magnetic flux can be calculated as:

$$\Phi = \int \vec{B} \cdot d\vec{S} = \int d\chi d\eta. \tag{5.68}$$

In the case that the surface integral is performed inside a load contour we obtain:

$$\Phi(\lambda) = \int_\lambda d\chi d\eta = \chi \int_\lambda d\eta = \begin{cases} \chi[\eta] \\ \chi(\eta_{max} - \eta_{min}) \end{cases} \tag{5.69}$$

There are two cases involved; in one case the load surfaces are topological cylinders, in this case η is not single valued and hence we obtain the upper value for $\Phi(\lambda)$. In a second case the load surfaces are topological spheres, in this case η is single valued and has minimal η_{min} and maximal η_{max} values. Hence the lower value of $\Phi(\lambda)$ is obtained. For example in some cases η is identical to twice the latitude angle θ. In those cases $\eta_{min} = 0$ (value at the "north pole") and $\eta_{max} = 2\pi$ (value at the "south pole").

Comparing the above equation with equation (5.62) we derive that η can be either **single valued** or **not single valued** and that its discontinuity across its cut in the non single valued case is $[\eta] = 2\pi$.

We will now show how the potentials α, β, ν can be derived. Let us calculate the vorticity $\vec{\omega}$ of the flow. By taking the curl of equation (5.27) we obtain:

$$\vec{\omega} \equiv \vec{\nabla} \times \vec{v} = \vec{\nabla}\alpha \times \vec{\nabla}\chi + \vec{\nabla}\beta \times \vec{\nabla}\eta. \tag{5.70}$$

The following identities are derived:

$$\vec{\omega} \cdot \vec{\nabla}\chi = (\vec{\nabla}\beta \times \vec{\nabla}\eta) \cdot \vec{\nabla}\chi = -\vec{\nabla}\beta \cdot \vec{B}, \tag{5.71}$$

$$\vec{\omega} \cdot \vec{\nabla}\eta = (\vec{\nabla}\alpha \times \vec{\nabla}\chi) \cdot \vec{\nabla}\eta = \vec{\nabla}\alpha \cdot \vec{B}. \tag{5.72}$$

Now let us perform integrations along $\vec{B}$ lines starting from an arbitrary point denoted as i to another arbitrary point denoted as r.

$$\beta(r) = -\int_i^r \frac{\vec{\omega} \cdot \vec{\nabla}\chi}{B} dl + \beta(i), \tag{5.73}$$

$$\alpha(r) = \int_i^r \frac{\vec{\omega} \cdot \vec{\nabla}\eta}{B} dl + \alpha(i). \tag{5.74}$$

The numbers $\alpha(i), \beta(i)$ can be chosen in an arbitrary way for each magnetic field line. Hence we have derived (in a non-unique way) the values of the α, β functions. Finally we can use equation (5.27) to derive the function ν for any point s within the flow:

$$\nu(s) = \int_i^s (\vec{v} - \alpha\vec{\nabla}\chi - \beta\vec{\nabla}\eta) \cdot d\vec{r} + \nu(i), \tag{5.75}$$

in which i is any arbitrary point within the flow, the result will not depend on the trajectory taken in the case that ν is single valued. If ν is not single valued on should introduce a cut which the integration trajectory should not cross.

5.4.5 Stationary barotropic magnetohydrodynamics

Stationary flows are a unique phenomena of Eulerian fluid dynamics which has no counter part in Lagrangian fluid dynamics. The stationary flow is defined by the fact that the physical fields $\vec{v}, \vec{B}, \rho$ do not depend on the temporal coordinate. This, however, does not imply that the corresponding potentials $\alpha, \beta, \chi, \eta, \nu$ are all functions of spatial coordinates alone. Moreover, it can be shown that choosing the potentials in such a way will lead to erroneous results in the sense that the stationary equations of motion can not be derived from the Lagrangian density $\hat{\mathcal{L}}$ given in equation (5.37). However, this problem can be amended easily as follows. Let us choose α, β, χ, ν to depend on the spatial coordinates alone. Let us choose η such that:

$$\eta = \bar{\eta} - t, \tag{5.76}$$

in which $\bar{\eta}$ is a function of the spatial coordinates. The Lagrangian density $\hat{\mathcal{L}}$ given in equation (5.37) will take the form:

$$\hat{\mathcal{L}} = \rho(\beta - \varepsilon(\rho)) - \frac{1}{2}\rho(\vec{\nabla}\nu + \alpha\vec{\nabla}\chi + \beta\vec{\nabla}\bar{\eta})^2 - \frac{1}{8\pi}(\vec{\nabla}\chi \times \vec{\nabla}\bar{\eta})^2. \tag{5.77}$$

The above functional can be compared with Vladimirov and Moffatt [7] equation 6.12 for incompressible flows in which their I is analogue to our β. Notice however, that while β is not a conserved quantity I is.

Varying the Lagrangian $\hat{L} = \int \hat{\mathcal{L}} d^3x$ with respect to $\nu, \alpha, \beta, \chi, \eta, \rho$ leads to the following equations:

$$\begin{aligned}
&\vec{\nabla} \cdot (\rho\hat{\vec{v}}) = 0, \\
&\rho\hat{\vec{v}} \cdot \vec{\nabla}\chi = 0, \\
&\rho(\hat{\vec{v}} \cdot \vec{\nabla}\bar{\eta} - 1) = 0, \\
&\hat{\vec{v}} \cdot \vec{\nabla}\alpha = \frac{\vec{\nabla}\bar{\eta} \cdot \hat{\vec{J}}}{\rho}, \\
&\hat{\vec{v}} \cdot \vec{\nabla}\beta = -\frac{\vec{\nabla}\chi \cdot \hat{\vec{J}}}{\rho}, \\
&\beta = \frac{1}{2}\hat{\vec{v}}^2 + w.
\end{aligned} \tag{5.78}$$

Calculations similar to the ones done in previous subsections will show that those equations lead to the stationary barotropic magnetohydrodynamic equations:

$$\vec{\nabla} \times (\hat{\vec{v}} \times \hat{\vec{B}}) = 0, \tag{5.79}$$

$$\rho(\hat{\vec{v}} \cdot \vec{\nabla})\hat{\vec{v}} = -\vec{\nabla}p(\rho) + \frac{(\vec{\nabla} \times \hat{\vec{B}}) \times \hat{\vec{B}}}{4\pi}. \tag{5.80}$$

5.5 The Simplified Hamiltonian Formalism

Let us derive the conjugate momenta of the variables appearing in the Lagrangian density $\hat{\mathcal{L}}$ defined in equation (5.37). A simple calculation will yield:

$$\pi_\nu \equiv \frac{\partial\hat{\mathcal{L}}}{\partial\left(\frac{\partial\nu}{\partial t}\right)} = -\rho, \quad \pi_\chi \equiv \frac{\partial\hat{\mathcal{L}}}{\partial\left(\frac{\partial\chi}{\partial t}\right)} = -\rho\alpha, \quad \pi_\eta \equiv \frac{\partial\hat{\mathcal{L}}}{\partial\left(\frac{\partial\eta}{\partial t}\right)} = -\rho\beta. \tag{5.81}$$

The rest of the canonical momenta $\pi_\rho, \pi_\alpha, \pi_\beta$ are null. It thus seems that the six functions appearing in the Lagrangian density $\hat{\mathcal{L}}$ can be divided to "approximate" conjugate pairs: $(\nu, \rho), (\chi, \alpha), (\eta, \beta)$. The Hamiltonian density $\hat{\mathcal{H}}$ can be now calculated as follows:

$$\hat{\mathcal{H}} = \pi_\nu \frac{\partial \nu}{\partial t} + \pi_\chi \frac{\partial \chi}{\partial t} + \pi_\eta \frac{\partial \eta}{\partial t} - \hat{\mathcal{L}} = \rho \left[\varepsilon(\rho) + \frac{1}{2} \hat{\vec{v}}^2 \right] + \frac{1}{8\pi} \hat{\vec{B}}^2, \tag{5.82}$$

in which $\hat{\vec{v}}$ is defined in equation (5.27) and $\hat{\vec{B}}$ is defined in equation (5.25). This Hamiltonian was previously introduced by Frenkel, Levich & Stilman [16] using somewhat different variables[3]. The equations derived from the above Hamiltonian density are similar to equations (5.24), equation (5.29) and equations (5.32) and will not be re-derived here. While Frenkel, Levich & Stilman [16] have postulated the Hamiltonian density appearing in equation (5.82), this Hamiltonian is here derived from a Lagrangian.

5.6 Simplified variational principle of stationary barotropic magnetohydrodynamics

In the previous section we have shown that barotropic magnetohydrodynamics can be described in terms of six first order differential equations or of an action principle from which those equations can be derived. This formalism was shown to apply to both stationary and non-stationary magnetohydrodynamics. For non-stationary magnetohydrodynamics we have shown that the number of functions can be further reduced to four, for stationary barotropic magnetohydrodynamics the situation can be improved even further. In fact we will show that for stationary barotropic magnetohydrodynamics three functions will suffice.

Consider equation (5.63), for a stationary flow it takes the form:

$$\vec{v} \cdot \vec{\nabla} \chi = 0. \tag{5.83}$$

Hence $\vec{v}$ can take the form:

$$\vec{v} = \frac{\vec{\nabla} \chi \times \vec{K}}{\rho}. \tag{5.84}$$

[3]The following notations are used in [16]: $\lambda = \chi, \Lambda = \eta, \mu = \pi_\chi, M = \pi_\eta, \phi = \nu$

However, the velocity field must satisfy the stationary mass conservation equation (5.3):

$$\vec{\nabla} \cdot (\rho \vec{v}) = 0. \tag{5.85}$$

We see that a sufficient condition (although not necessary) for $\vec{v}$ to solve equation (5.85) is that $\vec{K}$ takes the form $\vec{K} = \vec{\nabla} N$, where N is an arbitrary function. Thus, $\vec{v}$ may take the form:

$$\vec{v} = \frac{\vec{\nabla}\chi \times \vec{\nabla} N}{\rho}. \tag{5.86}$$

Let us now calculate $\vec{v} \times \vec{B}$ in which $\vec{B}$ is given by Sakurai's presentation equation (5.25):

$$\begin{aligned} \vec{v} \times \vec{B} &= \left(\frac{\vec{\nabla}\chi \times \vec{\nabla} N}{\rho}\right) \times (\vec{\nabla}\chi \times \vec{\nabla}\eta) \\ &= \frac{1}{\rho}\vec{\nabla}\chi(\vec{\nabla}\chi \times \vec{\nabla} N) \cdot \vec{\nabla}\eta. \end{aligned} \tag{5.87}$$

Since the flow is stationary N can be at most a function of the three comoving coordinates $\chi, \mu, \bar{\eta}$ defined in subsection 5.4.4, hence:

$$\vec{\nabla} N = \frac{\partial N}{\partial \chi}\vec{\nabla}\chi + \frac{\partial N}{\partial \mu}\vec{\nabla}\mu + \frac{\partial N}{\partial \bar{\eta}}\vec{\nabla}\bar{\eta}. \tag{5.88}$$

Inserting equation (5.88) into equation (5.87) will yield:

$$\vec{v} \times \vec{B} = \frac{1}{\rho}\vec{\nabla}\chi\frac{\partial N}{\partial \mu}(\vec{\nabla}\chi \times \vec{\nabla}\mu) \cdot \vec{\nabla}\eta. \tag{5.89}$$

Rearranging terms and using Sakurai's presentation equation (5.25) we can simplify the above equation and obtain:

$$\vec{v} \times \vec{B} = -\frac{1}{\rho}\vec{\nabla}\chi\frac{\partial N}{\partial \mu}(\vec{\nabla}\mu \cdot \vec{B}). \tag{5.90}$$

However, using equation (5.61) this will simplify to the form:

$$\vec{v} \times \vec{B} = -\vec{\nabla}\chi\frac{\partial N}{\partial \mu}. \tag{5.91}$$

Now let us consider equation (5.1); for stationary flows this will take the form:

$$\vec{\nabla} \times (\vec{v} \times \vec{B}) = 0. \tag{5.92}$$

Inserting equation (5.90) into equation (5.79) will lead to the equation:

$$\vec{\nabla}(\frac{\partial N}{\partial \mu}) \times \vec{\nabla}\chi = 0. \tag{5.93}$$

However, since N is at most a function of $\chi, \mu, \bar{\eta}$ it follows that $\frac{\partial N}{\partial \mu}$ is some function of χ:

$$\frac{\partial N}{\partial \mu} = -F(\chi). \tag{5.94}$$

This can be easily integrated to yield:

$$N = -\mu F(\chi) + G(\chi, \bar{\eta}). \tag{5.95}$$

Inserting this back into equation (5.86) will yield:

$$\vec{v} = \frac{\vec{\nabla}\chi \times (-F(\chi)\vec{\nabla}\mu + \frac{\partial G}{\partial \bar{\eta}}\vec{\nabla}\bar{\eta})}{\rho}. \tag{5.96}$$

Let us now replace the set of variables $\chi, \bar{\eta}$ with a new set $\chi', \bar{\eta}'$ such that:

$$\chi' = \int F(\chi) d\chi, \qquad \bar{\eta}' = \frac{\bar{\eta}}{F(\chi)}. \tag{5.97}$$

This will not have any effect on the Sakurai representation given in equation (5.25) since:

$$\vec{B} = \vec{\nabla}\chi \times \vec{\nabla}\eta = \vec{\nabla}\chi \times \vec{\nabla}\bar{\eta} = \vec{\nabla}\chi' \times \vec{\nabla}\bar{\eta}'. \tag{5.98}$$

However, the velocity will have a simpler representation and will take the form:

$$\vec{v} = \frac{\vec{\nabla}\chi' \times \vec{\nabla}(-\mu + G'(\chi', \bar{\eta}'))}{\rho}, \tag{5.99}$$

in which $G' = \frac{G}{F}$. At this point one should remember that μ was defined in equation (5.59) up to an arbitrary constant which can vary between magnetic field lines. Since the lines are labelled by their $\chi', \bar{\eta}'$ values it follows that we can add an arbitrary function of $\chi', \bar{\eta}'$ to μ without effecting its properties. Hence we can define a new μ' such that:

$$\mu' = \mu - G'(\chi', \bar{\eta}'). \tag{5.100}$$

Notice that μ' can be multi-valued; this will be discussed in somewhat more detail in subsection 5.6.3. Inserting equation (5.100) into equation (5.99) will lead to a simplified equation for $\vec{v}$:

$$\vec{v} = \frac{\vec{\nabla}\mu' \times \vec{\nabla}\chi'}{\rho}. \tag{5.101}$$

In the following the primes on $\chi, \mu, \bar{\eta}$ will be ignored. The above equation is analogous to Vladimirov and Moffatt's [7] equation 7.11 for incompressible flows, in which our μ and χ play the part of their A and Ψ. It is obvious that $\vec{v}$ satisfies the following set of equations:

$$\vec{v} \cdot \vec{\nabla}\mu = 0, \qquad \vec{v} \cdot \vec{\nabla}\chi = 0, \qquad \vec{v} \cdot \vec{\nabla}\bar{\eta} = 1, \tag{5.102}$$

to derive the right hand equation we have used both equation (5.60) and equation (5.25). Hence μ, χ are both comoving and stationary. As for $\bar{\eta}$ it satisfies the same equation as $\bar{\eta}$ defined in equation (5.76) as can be seen from equation (5.78). It can be easily seen that if:

$$basis = (\vec{\nabla}\chi, \vec{\nabla}\bar{\eta}, \vec{\nabla}\mu), \tag{5.103}$$

is a local vector basis at any point in space than their exists a dual basis:

$$dual\ basis = \frac{1}{\rho}(\vec{\nabla}\bar{\eta} \times \vec{\nabla}\mu, \vec{\nabla}\mu \times \vec{\nabla}\chi, \vec{\nabla}\chi \times \vec{\nabla}\bar{\eta}) = (\frac{\vec{\nabla}\bar{\eta} \times \vec{\nabla}\mu}{\rho}, \vec{v}, \frac{\vec{B}}{\rho}). \tag{5.104}$$

Such that:

$$basis_i \cdot dual\ basis_j = \delta_{ij}, \qquad i,j \in [1,2,3], \tag{5.105}$$

in which δ_{ij} is Kronecker's delta. Hence while the surfaces $\chi, \mu, \bar{\eta}$ generate a local vector basis for space, the physical fields of interest $\vec{v}, \vec{B}$ are part of the dual basis. By vector multiplying $\vec{v}$ and $\vec{B}$ and using equations (5.101,5.25) we obtain:

$$\vec{v} \times \vec{B} = \vec{\nabla}\chi, \tag{5.106}$$

this means that both $\vec{v}$ and $\vec{B}$ lie on χ surfaces and provide a vector basis for this two dimensional surface. The above equation can be compared with Vladimirov and Moffatt [7] equation 5.6 for incompressible flows in which their J is analogue to our χ.

5.6.1 The action principle

In the previous subsection we have shown that if the velocity field $\vec{v}$ is given by equation (5.101) and the magnetic field $\vec{B}$ is given by the Sakurai representation equation (5.25) than equations (5.1,5.2,5.3) are satisfied automatically for stationary flows. To complete the set of equations we will show how the Euler equations (5.4) can be derived from the action given in equation (5.6) in which both $\vec{v}$ and $\vec{B}$ are given by equation (5.101) and equation (5.25) respectively and the density ρ is given by equation (5.60):

$$\rho = \vec{\nabla}\mu \cdot \vec{B} = \vec{\nabla}\mu \cdot (\vec{\nabla}\chi \times \vec{\nabla}\eta) = \frac{\partial(\chi, \eta, \mu)}{\partial(x, y, z)}. \tag{5.107}$$

In this case the Lagrangian densityof equation (5.6) will take the form:

$$\mathcal{L} = \rho(\frac{1}{2}(\frac{\vec{\nabla}\mu \times \vec{\nabla}\chi}{\rho})^2 - \varepsilon(\rho)) - \frac{(\vec{\nabla}\chi \times \vec{\nabla}\eta)^2}{8\pi} \tag{5.108}$$

and can be seen explicitly to depend on only three functions. Let us make arbitrary small variations $\delta\alpha_i = (\delta\chi, \delta\eta, \delta\mu)$ of the functions $\alpha_i = (\chi, \eta, \mu)$. Let us define a Δ variation that does not modify the α_i's, such that:

$$\Delta\alpha_i = \delta\alpha_i + (\vec{\xi} \cdot \vec{\nabla})\alpha_i = 0, \tag{5.109}$$

in which $\vec{\xi}$ is the Lagrangian displacement, thus:

$$\delta\alpha_i = -\vec{\nabla}\alpha_i \cdot \vec{\xi}. \tag{5.110}$$

Which will lead to the equation:

$$\vec{\xi} \equiv -\frac{\partial\vec{r}}{\partial\alpha_i}\delta\alpha_i. \tag{5.111}$$

Making a variation of ρ given in equation (5.107) with respect to α_i will yield equation (5.8). Furthermore, taking the variation of $\vec{B}$ given by Sakurai's representation (5.25) with respect to α_i will yield equation (5.9). It remains to calculate $\delta\vec{v}$ by varying equation (5.101) this will yield:

$$\delta\vec{v} = -\frac{\delta\rho}{\rho}\vec{v} + \frac{1}{\rho}\vec{\nabla} \times (\rho\vec{\xi} \times \vec{v}). \tag{5.112}$$

Inserting equations (5.8,5.9,5.112) into equation (5.7) will yield:

$$\begin{aligned} \delta \mathcal{L} &= \vec{v} \cdot \vec{\nabla} \times (\rho \vec{\xi} \times \vec{v}) - \frac{\vec{B} \cdot \vec{\nabla} \times (\vec{\xi} \times \vec{B})}{4\pi} - \delta \rho (\frac{1}{2} \vec{v}^2 + w) \\ &= \vec{v} \cdot \vec{\nabla} \times (\rho \vec{\xi} \times \vec{v}) - \frac{\vec{B} \cdot \vec{\nabla} \times (\vec{\xi} \times \vec{B})}{4\pi} + \vec{\nabla} \cdot (\rho \vec{\xi})(\frac{1}{2} \vec{v}^2 + w) \end{aligned} \quad (5.113)$$

Using the well known vector identity:

$$\vec{A} \cdot \vec{\nabla} \times (\vec{C} \times \vec{A}) = \vec{\nabla} \cdot ((\vec{C} \times \vec{A}) \times \vec{A}) + (\vec{C} \times \vec{A}) \cdot \vec{\nabla} \times \vec{A} \quad (5.114)$$

and the theorem of Gauss we can write now equation (5.7) in the form:

$$\begin{aligned} \delta A &= \int dt \{ \oint d\vec{S} \cdot [\rho (\vec{\xi} \times \vec{v}) \times \vec{v} - \frac{(\vec{\xi} \times \vec{B}) \times \vec{B}}{4\pi} + (\frac{1}{2} \vec{v}^2 + w) \rho \vec{\xi}] \\ &+ \int d^3 x \vec{\xi} \cdot [\rho \vec{v} \times \vec{\omega} + \vec{J} \times \vec{B} - \rho \vec{\nabla} (\frac{1}{2} \vec{v}^2 + w)] \}. \end{aligned} \quad (5.115)$$

The time integration is of course redundant in the above expression. Also notice that we have used the current definition equation (5.5) and the vorticity definition equation (5.70). Suppose now that $\delta A = 0$ for a $\vec{\xi}$ such that the boundary term (including both the boundary of the domain and relevant cuts) in the above equation is null but that $\vec{\xi}$ is otherwise arbitrary, then it entails the equation:

$$\rho \vec{v} \times \vec{\omega} + \vec{J} \times \vec{B} - \rho \vec{\nabla} (\frac{1}{2} \vec{v}^2 + w) = 0. \quad (5.116)$$

Using the well known vector identity:

$$\frac{1}{2} \vec{\nabla} (\vec{v}^2) = (\vec{v} \cdot \vec{\nabla}) \vec{v} + \vec{v} \times (\vec{\nabla} \times \vec{v}) \quad (5.117)$$

and rearranging terms we recover the stationary Euler equation:

$$\rho (\vec{v} \cdot \vec{\nabla}) \vec{v} = -\vec{\nabla} p + \vec{J} \times \vec{B}. \quad (5.118)$$

5.6.2 The case of an axi-symmetric magnetic field

Consider an axi-symmetric magnetic field such that the magnetic field is dependent only on the coordinate R which is the distance from the axis of symmetry and the coordinate z which is the distance along the axis of symmetry from an arbitrary origin on the axis. Thus:

$$\vec{B} = \vec{B}(R, z). \tag{5.119}$$

Any axi-symmetric magnetic field satisfying equation (5.2) can be represented in the form:

$$\vec{B} = \vec{\nabla} P \times \vec{\nabla}(\frac{\phi}{2\pi}) + 2\pi R B_\phi \vec{\nabla}(\frac{\phi}{2\pi}). \tag{5.120}$$

In which ϕ is the azimuthal angle defined in the conventional way and B_ϕ is the component of $\vec{B}$ in the ϕ direction. The function $P = P(R, z)$ is the flux through a circle of radius R at height z:

$$P(R, z) = \int_{(R,z)} \vec{B} \cdot d\vec{S} = 2\pi \int_0^R B_z(R', z) R' dR'. \tag{5.121}$$

For finite field configurations P will have a maximum $P_m = P(R_m, z_m)$ at some R_m, z_m. This circle $R = R_m$ will form a line toroid with the other constant P surfaces nearby forming a nested set. There can be several such local maxima with local nested toroids in a general configuration but the simpler case has just one.

Let us study the relations between the functions P, B_ϕ and the functions χ, η given in equation (5.17). Assuming that the density ρ is axi-symmetric one can see the magnetic load defined in equation (5.56) is also axi-symmetric and that the surfaces of constant load are surfaces of revolution around the axis of symmetry. From equation (5.62) we deduce that $\chi = \chi(R, z)$. Expressing equation (5.17) in terms of the coordinates R, ϕ, z results in:

$$\vec{B} = (-\frac{1}{R}\partial_z \chi \partial_\phi \eta)\hat{R} + (\partial_z \chi \partial_R \eta - \partial_R \chi \partial_z \eta)\hat{\phi} + (\frac{1}{R}\partial_R \chi \partial_\phi \eta)\hat{z}. \tag{5.122}$$

In which ∂_y is a short hand notation for $\frac{\partial}{\partial y}$ and $\hat{y}$ is a unit vector perpendicular to the constant y surface. Comparing equation (5.122) with equation (5.120) we arrive with the set of equations:

$$\partial_z P = \partial_z \chi \partial_\phi (2\pi\eta), \qquad \partial_R P = \partial_R \chi \partial_\phi (2\pi\eta). \tag{5.123}$$

From which we derive the equation:

$$\partial_z P \partial_R \chi - \partial_R P \partial_z \chi = 0 \Rightarrow \vec{\nabla} P \times \vec{\nabla} \chi = 0. \tag{5.124}$$

Hence:

$$P = F(\chi), \tag{5.125}$$

where F is an arbitrary function. We deduce that P is just another type of labelling of the load surfaces. Thus equation (5.123) will lead to:

$$\partial_\phi(2\pi\eta) = \frac{dP}{d\chi} \Rightarrow \eta = \frac{\phi}{2\pi}\frac{dP}{d\chi} + \tilde{\eta}(R, z). \tag{5.126}$$

This should be compared with the result of Young et al. [13, equation 5.1]. Substituting the above result in equation (5.122) will lead to the equation:

$$B_\phi \hat{\phi} = (\partial_z \chi \partial_R \tilde{\eta} - \partial_R \chi \partial_z \tilde{\eta})\hat{\phi} = \vec{\nabla}\chi \times \vec{\nabla}\tilde{\eta}. \tag{5.127}$$

This can also be written as:

$$B_\phi = \hat{\phi} \cdot (\vec{\nabla}\chi \times \vec{\nabla}\tilde{\eta}) = (\hat{\phi} \times \vec{\nabla}\chi) \cdot \vec{\nabla}\tilde{\eta}. \tag{5.128}$$

Hence B_ϕ is proportional to the gradient of $\tilde{\eta}$ along the $\hat{\phi} \times \vec{\nabla}\chi$ direction. Since $\hat{\phi} \times \vec{\nabla}\chi$ is known we can integrate along this vector to obtain a non-unique solution for $\tilde{\eta}$:

$$\tilde{\eta} = \int \frac{B_\phi}{|\hat{\phi} \times \vec{\nabla}\chi|} dl, \tag{5.129}$$

in which dl is a line element along the $\hat{\phi} \times \vec{\nabla}\chi$ line.

5.6.3 The case of a magnetic field on a toroid

Our previous definitions of the surfaces of constant load given in equation (5.56) is ambiguous when the field lines are "surface filling" eg on a toroid and give no result when the field lines are "volume filling". At equilibrium $\vec{B}$ and $\vec{v}$ lie in surfaces (there is an exception when $\vec{B}$ and $\vec{v}$ are parallel and fill volumes). Our former considerations apply unchanged if these surfaces have the topology of cylinders but they need generalization when the surfaces have the topology of toroids nested on a line (a similar discussion in which non-magnetic fluids are considered can be found in [27]). We consider a surface Σ spanning that line toroid. Each toroid T will meet Σ in a loop. Consider the magnetic flux $\Phi(T)$ through that part of Σ within the loop and the mass enclosed by the toroid $m(T)$. The the mass outside the toroid is $\tilde{m}(T) = M - m(T)$. Now express $\tilde{m}$ as a function $\tilde{m}(\Phi)$ of the magnetic flux Φ, then a definition of magnetic load analogous to that for "cylinders" parallel to the axis is:

$$\lambda = \frac{d\tilde{m}}{d\Phi}. \tag{5.130}$$

However, there are now two loads corresponding to the two fluxes associated with a given toroid. The other load is obtained by taking a cut across the "short" circle section of the torus say of constant ϕ. The magnetic flux Φ^* through such a cross section may be expressed as a function of the total mass $m(T)$ within the toroid and

$$\lambda^* = \frac{dm}{d\Phi^*}, \tag{5.131}$$

is a second different load. Of course it is also permissible to reexpress the flux Φ^* as a function of the flux Φ then we find:

$$\lambda^* = \frac{dm}{d\Phi^*} = \frac{\frac{dm}{d\Phi}}{\frac{d\Phi^*}{d\Phi}} = -\frac{\lambda}{\frac{d\Phi^*}{d\Phi}}. \tag{5.132}$$

The surfaces of constant λ are of course the toroids T which also have λ^* constant.

A similar problem may arise with the definition of the magnetic metage defined in equation (5.59). We may wish to define this quantity using the magnetic field $\vec{B}$ and velocity field $\vec{v}$. Since those vectors provide a vector basis on the load surface, they can be combined in such a way say: $\vec{B}+\gamma(\chi)\vec{v}$ to create a vector which is directed along the large loop of the toroid. (A different γ will leave only twists around the short way.) This combination represents an unwinding of the field lines so that they no longer twist around the short (long) way. Those loops can be thought as composing the surface Σ. Another surface Σ' also composed of such untwisted loops can be so chosen that the mass between Σ and Σ' and between loads λ and $\lambda + d\lambda$ is some fixed fraction of $\frac{dm}{d\lambda}d\lambda$. Such Σ' form suitable constant metage surfaces μ corresponding to partial loads λ. Notice that $2\pi\mu$ then describes the angle from Σ turned around the toroid by the short way to reach any chosen point. Similar use of the other load λ^* allows us to define another generalized angle μ^* measured around the long way. A somewhat less physical approach is given below.

Let us consider a toroid of constant magnetic load. Dungey [25, p. 31] has considered the case in which magnetic field lines lie on a torus. He has shown that one of the functions (ie η) involved in the representation (5.17) should be non-single valued and therefore a cut should be introduced.

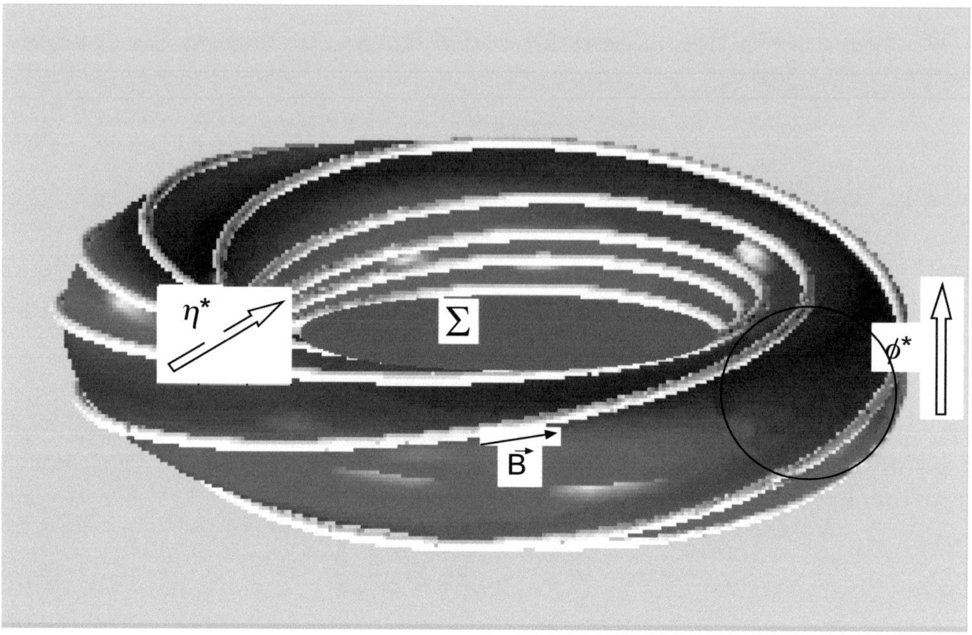

Figure 5.3: A torus of magnetic field lines

In order to obtain a simple looking cuts we will replace the previous set of functions μ, η with a new set ϕ^*, η^*, which will be defined as follows:

$$\phi^* \equiv \frac{\mu + G(\chi)\eta}{\Omega(\chi)},$$
$$\eta^* \equiv \frac{\eta - \phi^*}{f(\chi)}, \tag{5.133}$$

G, Ω, f are arbitrary functions of χ. Therefore μ and η can be given in terms of ϕ^*, η^* as:

$$\eta = \phi^* + f(\chi)\eta^*,$$
$$\mu = -G(\chi)\eta + \Omega(\chi)\phi^*, \tag{5.134}$$

η^* can be considered as an angle varying over the small circle of the torus, while ϕ^* can be considered as an angle varying over the large circle of the torus as in figure (5.3). On the torus of constant magnetic load the ϕ^*, η^* functions have a simple "cut" structure. The above equation can also serve as a "definition" of μ. Inserting equations (5.134) into equation (5.17) and equation (5.101) will result in the following set of equations:

$$\begin{aligned}\vec{B} &= \vec{\nabla}\chi \times \vec{\nabla}\phi^* + f(\chi)\vec{\nabla}\chi \times \vec{\nabla}\eta^*, \\ \vec{v} &= G(\chi)\frac{\vec{B}}{\rho} + \Omega(\chi)\frac{\vec{\nabla}\phi^* \times \vec{\nabla}\chi}{\rho}.\end{aligned} \tag{5.135}$$

Hence $\vec{B}$ is partitioned into two vectors circulating along the small and large circles of the torus. While $\vec{v}$ has two vector components one along the magnetic field $\vec{B}$ and another along the small circle.

5.7 Topological constants of motion

Magnetohydrodynamics is known to have the following two topological constants of motion; one is the magnetic helicity:

$$\mathcal{H}_M \equiv \int \vec{B} \cdot \vec{A} d^3x, \tag{5.136}$$

which is known to measure the degree of knottiness of lines of the magnetic field $\vec{B}$ [28]. The domain of integration in equation (5.136) is the entire space, obviously regions containing a null magnetic field will have a null contribution to the integral. In the above equation $\vec{A}$ is the vector potential defined implicitly by the equation:

$$\vec{B} = \vec{\nabla} \times \vec{A}. \tag{5.137}$$

The other topological constant is the magnetic cross helicity:

$$\mathcal{H}_C \equiv \int \vec{B} \cdot \vec{v} d^3x, \tag{5.138}$$

characterizing the degree of cross knottiness of the magnetic field and velocity lines. The domain of integration in equation (5.138) is the magnetohydrodynamic flow domain.

5.7.1 Representation in terms of the magnetohydrodynamic potentials

Let us write the topological constants given in equation (5.136) and equation (5.138) in terms of the magnetohydrodynamic potentials $\alpha, \beta, \chi, \eta, \nu, \rho$ introduced in previous sections.

First let us combine equation (5.17) with equation (5.137) to obtain the equation:

$$\vec{\nabla} \times (\vec{A} - \chi \vec{\nabla} \eta) = 0, \tag{5.139}$$

this leads immediately to the result:

$$\vec{A} = \chi \vec{\nabla} \eta + \vec{\nabla} \zeta, \tag{5.140}$$

in which ζ is some function. Let us now calculate the scalar product $\vec{B} \cdot \vec{A}$:

$$\vec{B} \cdot \vec{A} = (\vec{\nabla} \chi \times \vec{\nabla} \eta) \cdot \vec{\nabla} \zeta. \tag{5.141}$$

However, since we have the local vector basis: $(\vec{\nabla} \chi, \vec{\nabla} \eta, \vec{\nabla} \mu)$ we can write $\vec{\nabla} \zeta$ as:

$$\vec{\nabla} \zeta = \frac{\partial \zeta}{\partial \chi} \vec{\nabla} \chi + \frac{\partial \zeta}{\partial \mu} \vec{\nabla} \mu + \frac{\partial \zeta}{\partial \eta} \vec{\nabla} \eta. \tag{5.142}$$

Hence we can write:

$$\vec{B} \cdot \vec{A} = \frac{\partial \zeta}{\partial \mu} (\vec{\nabla} \chi \times \vec{\nabla} \eta) \cdot \vec{\nabla} \mu = \frac{\partial \zeta}{\partial \mu} \frac{\partial(\chi, \eta, \mu)}{\partial(x, y, z)}. \tag{5.143}$$

Let us think of the entire space outside the magnetohydrodynamic domain as containing low density matter in this case we can define the metage μ over the entire portion of space containing magnetic field lines and the integration domain of equation (5.136) and equation (5.138) coincide. Now we can insert equation (5.143) into equation (5.136) to obtain the expression:

$$\mathcal{H}_M = \int \frac{\partial \zeta}{\partial \mu} d\mu d\chi d\eta. \tag{5.144}$$

The reader should notice that in some scenarios it may be that the flow domain should be divided into patches in which different definitions of μ, χ, η apply to different domains, we do not see this as a limitation for our formalism since the topology of the flow is conserved by the equations of Magnetohydrodynamics. In those cases $\mathcal{H}_M$ should be calculated as sum of the contributions from each patch. We can think about the magnetohydrodynamic domain as composed of thin closed tubes of magnetic lines each labelled by (χ, η). Performing the integration along such a thin tube in the metage direction results in:

$$\oint_{\chi,\eta} \frac{\partial \zeta}{\partial \mu} d\mu = [\zeta]_{\chi,\eta}, \tag{5.145}$$

in which $[\zeta]_{\chi,\eta}$ is the discontinuity of the function ζ along its cut. Thus a thin tube of magnetic lines in which ζ is single valued does not contribute to the magnetic helicity integral. Inserting equation (5.145) into equation (5.144) will result in:

$$\mathcal{H}_M = \int [\zeta]_{\chi,\eta} d\chi d\eta = \int [\zeta] d\Phi, \tag{5.146}$$

in which we have used equation (5.68). Hence:

$$[\zeta] = \frac{d\mathcal{H}_M}{d\Phi}, \tag{5.147}$$

the discontinuity of ζ is thus the density of magnetic helicity per unit of magnetic flux in a tube. We deduce that the Sakurai representation does not entail zero magnetic helicity, rather it is perfectly consistent with non zero magnetic helicity as was demonstrated above and in agreement to the claims made by Frenkel, Levich & Stilman [16]. Notice however, that the topological structure of the magnetohydrodynamic flow constrain the gauge freedom which is usually attributed to vector potential $\vec{A}$ and limits it to single valued functions. Moreover, while the choice of $\vec{A}$ is arbitrary since one can add to $\vec{A}$ an arbitrary gradient of a single valued function which may lead to different choices of ζ the discontinuity value $[\zeta]$ is not arbitrary and has a physical meaning given above.

Let us now introduce the velocity expression given in equation (5.27) and calculate the scalar product of $\vec{B}$ and $\vec{v}$, using the same arguments as in the previous paragraph will lead to the expression:

$$\vec{v} \cdot \vec{B} = \frac{\partial \nu}{\partial \mu}(\vec{\nabla}\chi \times \vec{\nabla}\eta) \cdot \vec{\nabla}\mu = \frac{\partial \nu}{\partial \mu}\frac{\partial(\chi, \eta, \mu)}{\partial(x, y, z)}. \tag{5.148}$$

Inserting equation (5.148) into equation (5.138) will result in:

$$\mathcal{H}_C = \int \frac{\partial \nu}{\partial \mu} d\mu d\chi d\eta. \tag{5.149}$$

We can think about the magnetohydrodynamic domain as composed of thin closed tubes of magnetic lines each labelled by (χ, η). Performing the integration along such a thin tube in the metage direction results in:

$$\oint_{\chi,\eta} \frac{\partial \nu}{\partial \mu} d\mu = [\nu]_{\chi,\eta}, \tag{5.150}$$

in which $[\nu]_{\chi,\eta}$ is the discontinuity of the function ν along its cut. Thus a thin tube of magnetic lines in which ν is single valued does not contribute to the cross helicity integral. Inserting equation (5.150) into equation (5.149) will result in:

$$\mathcal{H}_C = \int [\nu]_{\chi,\eta} d\chi d\eta = \int [\nu] d\Phi, \tag{5.151}$$

in which we have used equation (5.68). Hence:

$$[\nu] = \frac{d\mathcal{H}_C}{d\Phi}, \tag{5.152}$$

the discontinuity of ν is thus the density of cross helicity per unit of magnetic flux. We deduce that a flow with null cross helicity will have a single valued ν function alternatively, a non single valued ν will entail a non zero cross helicity. Furthermore, from equation (5.29) it is obvious that:

$$\frac{d[\nu]}{dt} = 0. \tag{5.153}$$

We conclude that not only is the magnetic cross helicity conserved as an integral quantity of the entire magnetohydrodynamic domain but also the (local) density of cross helicity per unit of magnetic flux is a conserved quantity as well.

In the following sub section we give a simple example which will demonstrate some of the general assertions of this paragraph.

5.7.2 A Helical Stratified Magnetic Field

Consider a magnetohydrodynamic flow of uniform density ρ. Furthermore assume that the flow contains a helical stratified magnetic field:

$$\vec{B} = \begin{cases} 2B_\perp(1-\frac{R}{a})\hat{\phi} + B_{z0}\hat{z} & R < a \\ 0 & R > a \end{cases} \tag{5.154}$$

in which R, ϕ, z are the standard cylindrical coordinates, $\hat{R}, \hat{\phi}, \hat{z}$ are the corresponding unit vectors and $B_{z0}, B_\perp$ are constants. The magnetic field is contained in a cylinder of Radius a and is independent of z. A possible choice of the vector potential $\vec{A}$ is:

$$\vec{A} = \begin{cases} B_{z0}x\hat{y} + B_\perp a(1-\frac{R}{a})^2\hat{z}, & R < a \\ 0 & R > a \end{cases} \tag{5.155}$$

in which $\hat{y}$ is a unit vector in the y direction. Let us calculate the magnetic helicity of the field using equation (5.136) . In order to obtain a finite magnetic helicity we assume that the field is contained between the planes $z = 0$ and $z = 1$ furthermore we assume that the planes $z = 0$ and $z = 1$ can be identified such that the magnetic field lines are closed. Thus the domain becomes a topological torus. Inserting equation (5.154) and equation (5.155) into equation (5.136) will result in:

$$\mathcal{H}_M \equiv \int \vec{B} \cdot \vec{A} d^3x = \frac{\pi}{3} a^3 B_{z0} B_\perp \tag{5.156}$$

First let us calculate to load using equation (5.56) (we assume that $R < a$ in the following calculations) we obtain:

$$\lambda = \rho \frac{4\pi B_\perp (1 - \frac{R}{a}) R + B_{z0}}{B_{z0}^2 + 4B_\perp^2 (1 - \frac{R}{a})^2} = \lambda(R), \tag{5.157}$$

hence the load surfaces are cylinders. The χ function can now be calculated according to equation (5.62) to yield the value:

$$\chi = \frac{1}{2} B_{z0} R^2. \tag{5.158}$$

Solving equation (5.67) for η we obtain the following non unique solution:

$$\eta = \frac{2B_\perp}{B_{z0}} (1 - \frac{R}{a}) \frac{z}{a} + \phi. \tag{5.159}$$

Substituting equation (5.158), equation (5.159) and equation (5.155) into equation (5.140) we can solve for ζ and obtain:

$$\zeta = B_\perp z (a - R) + \frac{1}{2} B_{z0} x y. \tag{5.160}$$

Since we have identified the $z = 0$ and $z = 1$ planes the z coordinate is not single valued and therefore ζ is a non single valued function which has a discontinuity value:

$$[\zeta] = B_\perp (a - R). \tag{5.161}$$

Thus we can calculate the magnetic helicity using equation (5.146) and obtain:

$$\mathcal{H}_M = \int [\zeta] d\Phi = \frac{\pi}{3} a^3 B_{z0} B_\perp, \tag{5.162}$$

which coincides with the result of equation (5.144).

5.7.3 Self Knotted Magnetic Field Lines on Nested Tori

Consider a magnetohydrodynamic flow of uniform density ρ. Furthermore assume (following Moffat [28]) that the flow contains a vector potential:

$$\vec{A} = \vec{\nabla}\Psi \times \vec{\nabla}\phi + \alpha\Psi\vec{\nabla}\phi = \frac{1}{R}\partial_R\Psi\hat{z} - \frac{1}{R}\partial_z\Psi\hat{R} + \frac{\alpha\Psi}{R}\hat{\phi}, \qquad \vec{\nabla}\phi = \frac{\hat{\phi}}{R}, \tag{5.163}$$

in which as in the previous section R, ϕ, z are the standard cylindrical coordinates, $\hat{R}, \hat{\phi}, \hat{z}$ are the corresponding unit vectors, α is constant and $\Psi = \Psi(R, z)$ is an arbitrary function of R and z. The magnetic field can be calculated using equation (5.137) to be:

$$\vec{B} = \frac{\alpha}{R}\partial_R\Psi\hat{z} - \frac{\alpha}{R}\partial_z\Psi\hat{R} - \frac{D^2\Psi}{R}\hat{\phi}. \tag{5.164}$$

In which according to Moffat [28] the operator D^2 is defined as:

$$D^2 = \partial_z^2 + R\partial_R(\frac{1}{R}\partial_R). \tag{5.165}$$

Obviously both $\vec{A}$ and $\vec{B}$ lie on Ψ surfaces since:

$$\vec{\nabla}\Psi \cdot \vec{A} = \vec{\nabla}\Psi \cdot \vec{B} = 0. \tag{5.166}$$

Let us define the variable r:

$$r = \sqrt{z^2 + (R-1)^2}. \tag{5.167}$$

And let us assume that $\Psi = \Psi(r)$ is a function of r. In this case surfaces of constant Ψ are nested tori. The magnetic field is assumed to be confined between the tori $0 \leq r \leq a$ in which a is an arbitrary number such that $0 < a < 1$. A depiction of a $R - z$ cross section of the nested tori is given in figure 5.4 below. A typical field line of the magnetic field given in equation (5.164) is self knotted in the sense of Moffat [28] as is evident from figure 5.5.

Next, following section 5.6.3 we define two functions with simple cuts ϕ^* and η^*. In which η^* can be considered as an angle varying over the small circle of the torus, while ϕ^* can be considered as an angle varying over the

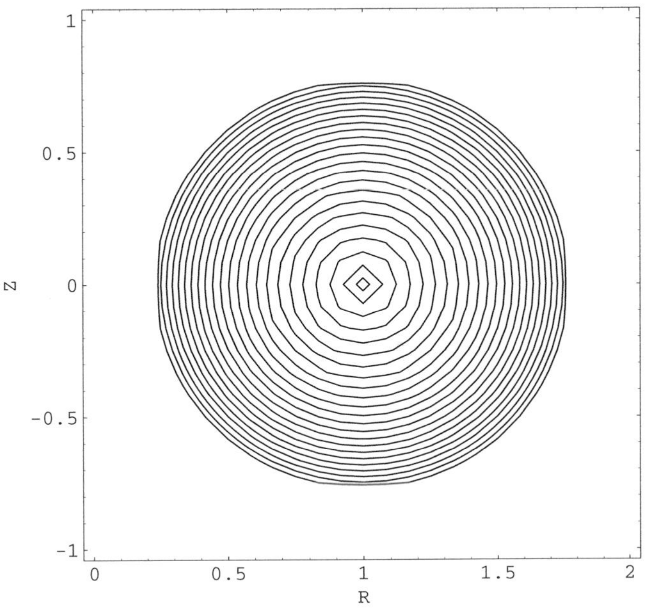

Figure 5.4: $R - z$ cross section of the nested tori.

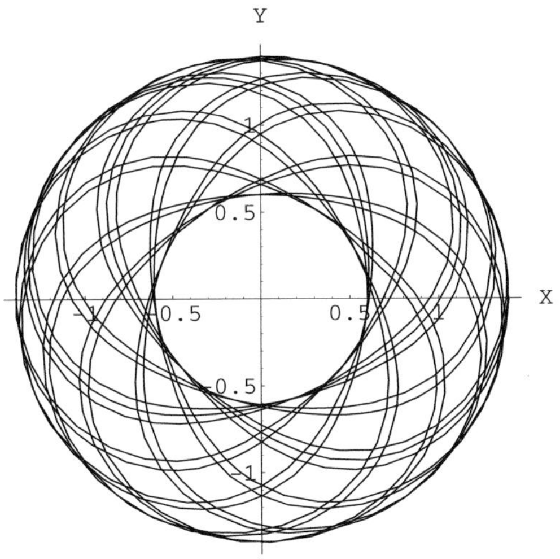

Figure 5.5: A numerically integrated field line assuming that $\Psi = r + r^3, \alpha = 1$ and starting from the point $R = 0.6, \phi = 0, z = 0$. The plot shows twenty rotations.

large circle of the torus. Hence $\phi^* = \phi$ is just the standard azimuthal angle and η^* can be defined as:

$$\eta^* = \arctan \frac{z}{R-1}. \tag{5.168}$$

Obviously the Ψ surfaces are also the λ surfaces. Therefore we can calculate χ using equation (5.62) were we calculate the magnetic flux into the surface between the degenerate torus $r = 0$ and any other torus given by some value of Ψ. There are two ways to do this but it seems that the simpler way is to take the surface which is perpendicular to $\hat{\eta^*}$ which is a unit vector in the η^* direction. Hence we obtain:

$$\chi = \frac{1}{2\pi}\int \vec{B}\cdot d\vec{S} = \frac{1}{2\pi}\oint \vec{A}\cdot d\vec{l} = \frac{1}{2\pi}\int_0^{2\pi} A_\phi R d\phi = A_\phi R = \alpha\Psi, \tag{5.169}$$

in the above we assumed that $\Psi(0) = 0$. Let us now calculate the function η by solving equation (5.67). It is easy to show that η is of the form:

$$\eta = \phi + C(z, R), \tag{5.170}$$

in which $C(z, R)$ is a solution of:

$$B_\phi = \partial_z\chi\partial_R C - \partial_z\chi\partial_z C. \tag{5.171}$$

Writing the above equation in terms of r, η^* coordinates we obtain:

$$\begin{aligned} -\frac{1}{1+r\cos\eta^*}\left(\Psi'' + \frac{1}{1+r\cos\eta^*}\frac{\Psi'}{r}\right) &= -\frac{\alpha\Psi'}{r}\partial_{\eta^*}C, \\ \Psi' \equiv \frac{d\Psi}{dr}, \quad \Psi'' = \frac{d^2\Psi}{dr^2} & \end{aligned} \tag{5.172}$$

C can be integrated to yield the solution:

$$C = \frac{1}{\alpha}\left[\frac{r\Psi''}{\Psi'}I(r,\eta^*) + II(r,\eta^*)\right], \tag{5.173}$$

in which:

$$\begin{aligned} I(r,\eta^*) &\equiv \int \frac{d\eta^*}{1+r\cos\eta^*} \\ &= \frac{2}{\sqrt{1-r^2}}\left[\arctan\left(\sqrt{\frac{1-r}{1+r}}\tan\left(\frac{\eta^*}{2}\right)\right) + \left\{\begin{array}{ll} 0, & 0 \le \eta^* < \pi \\ \pi, & \pi \le \eta^* < 2\pi. \end{array}\right.\right] \end{aligned} \tag{5.174}$$

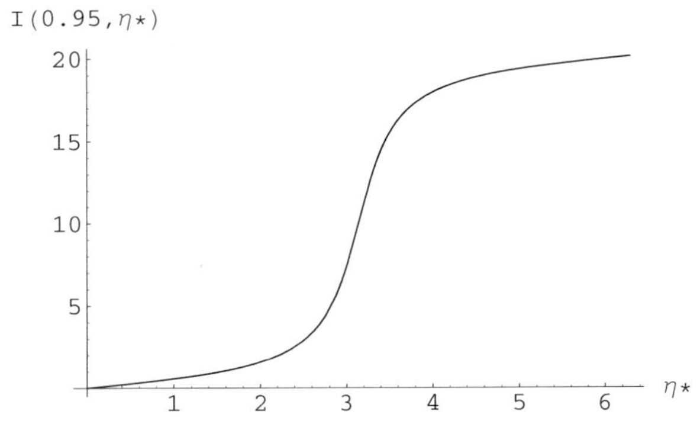

Figure 5.6: $I(r, \eta^*)$ for $r = 0.95$.

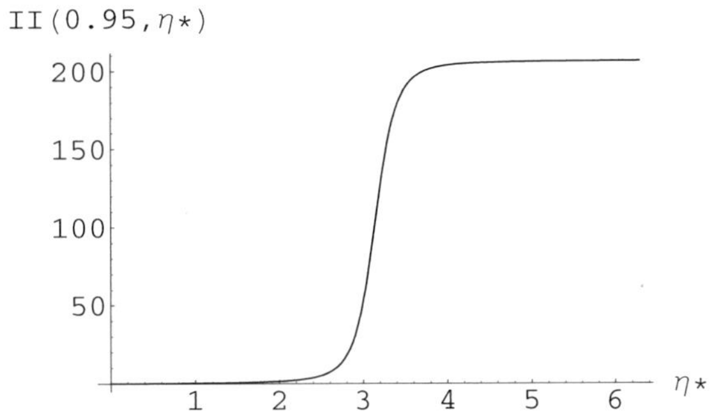

Figure 5.7: $II(r, \eta^*)$ for $r = 0.95$.

and

$$II(r, \eta^*) \equiv \int \frac{d\eta^*}{(1 + r\cos\eta^*)^2} = \frac{I(r, \eta^*)}{1 - r^2} - \frac{r\sin\eta^*}{(1 - r^2)(1 + r\cos\eta^*)}. \qquad (5.175)$$

Plots of $I(r, \eta^*)$ and $II(r, \eta^*)$ are given in figures 5.6 and 5.7 respectively.

Obviously $I(r, \eta^*)$ and $II(r, \eta^*)$ are non-single valued functions. Their discontinuity values across the cut are given by:

$$[I(r, \eta^*)] = \frac{2\pi}{\sqrt{1 - r^2}}, \qquad [I(r, \eta^*)] = \frac{2\pi}{(1 - r^2)^{3/2}}\,. \qquad (5.176)$$

Therefore $C(r, \eta^*)$ is also a non-single valued function. Using equation (5.173) we obtain the following discontinuity value of $C(r, \eta^*)$ across the cut:

$$[C] = \frac{2\pi}{\alpha\sqrt{1 - r^2}}\left(\frac{r\Psi''}{\Psi'} + \frac{1}{1 - r^2}\right). \qquad (5.177)$$

It remains to calculate the value of the function ζ, this can be done using equation (5.140). Inserting into equation (5.140) the value of η given in equation (5.170), we obtain:

$$\vec{A} = \chi\vec{\nabla}\eta + \vec{\nabla}\zeta = \frac{\alpha\Psi}{R}\hat{\phi} + \alpha\Psi\vec{\nabla}C + \vec{\nabla}\zeta. \tag{5.178}$$

Taking into account equation (5.163) in the above equation leads to:

$$\vec{\nabla}\zeta = \frac{1}{R}\partial_R\Psi\hat{z} - \frac{1}{R}\partial_z\Psi\hat{R} - \alpha\Psi\vec{\nabla}C. \tag{5.179}$$

The above equation implies that ζ is a function of R, z (or r, η^*) only. Writing equation (5.179) in terms of the r, η^* coordinates we arrive at a set of two equations:

$$\frac{1}{r}\partial_{\eta^*}\zeta = -\frac{\alpha\Psi}{r}\partial_{\eta^*}C + \frac{\Psi'}{1 + r\cos\eta^*}, \qquad \partial_r\zeta = -\alpha\Psi\partial_r C. \tag{5.180}$$

Solving equations (5.180) we arrive at the solution:

$$\zeta(r,\eta^*) = r\Psi' I(r,\eta^*) - \alpha\Psi C = rI(r,\eta^*)\left(\Psi' - \frac{\Psi\Psi''}{\Psi'}\right) - \Psi II(r,\eta^*) \tag{5.181}$$

Obviously $\zeta(r,\eta^*)$ is a non-single valued function with the following discontinuity value across the cut:

$$[\zeta(r,\eta^*)] = \frac{2\pi}{\sqrt{1-r^2}}\left(r(\Psi' - \frac{\Psi\Psi''}{\Psi'}) - \frac{\Psi}{1-r^2}\right) \tag{5.182}$$

Let us calculate the magnetic helicity of the field using equation (5.146), equation (5.182) and equation (5.169), we arrive at the result:

$$\begin{aligned} \mathcal{H}_M &= \int [\zeta] d\Phi = \int_0^a [\zeta] 2\pi\alpha\Psi' dr \\ &= (2\pi)^2\alpha \int_0^a \frac{dr}{\sqrt{1-r^2}}\left(r((\Psi')^2 - \Psi\Psi'') - \frac{\Psi\Psi'}{1-r^2}\right). \end{aligned} \tag{5.183}$$

A direct calculation using equation (5.136) will yield an identical result. This integral can be calculated either analytically or numerically for any reasonable function $\Psi(r)$. For example taking $\Psi(r) = r + r^3$ and $a = 0.9$ we calculated $\mathcal{H}_M$ numerically and obtained $\mathcal{H}_M = -4.6167(2\pi)^2\alpha$. To conclude we do not see any principle difficulty in calculating the functions defined in this work for self knotted magnetic field lines. It may be so that the functions should be derived numerically if the nested tori are distorted.

5.7.4 Cross Helicity Conservation via the Noether Theorem

The conservation of helicity $\int \vec{v} \cdot \vec{\omega} d^3x$ in ideal (non-magnetic) barotropic fluid when certain conditions are satisfied in particular, when $\vec{\omega} \cdot \vec{n} = 0$ on the (Lagrangian) surface bounding the volume of integration was discovered by Moffat [28]. Moreau [29] has discussed the conservation of helicity from the group theoretical point of view. In his paper he used an enlarged Arnold symmetry group [30] of fluid element labelling to generate the conservation of helicity. Yahalom [31, 32] has shown that the symmetry group generating conservation of helicity becomes a very simple one parameter translation group in the space of labels (alpha space) when represented by Lynden-Bell and Katz [33] labelling. We will now show that in the case of magnetohydrodynamics the same one parameter translation group will generate the magnetic cross helicity via the Noether Theorem.

Let us denote the initial position of a fluid element by (x_0^k), by mass conservation:

$$\rho(x^k)d^3x = \rho(x_0^k)d^3x_0 = \rho(x_0^k)\frac{\partial(x_0^1, x_0^2, x_0^3)}{\partial(\alpha^1, \alpha^2, \alpha^3)}d^3\alpha. \tag{5.184}$$

Since the initial position of a fluid element can not depend on time it must depend on the label only, and therefore by an appropriate choice of the $\alpha's$ we obtain:

$$\rho(x^k)d^3x = d^3\alpha, \qquad \rho = \frac{\partial(\alpha^1, \alpha^2, \alpha^3)}{\partial(x^1, x^2, x^3)}. \tag{5.185}$$

Where we assume that the above expressions of course exist. Let us look at the action A defined in equation (5.6).

From the discussion following equation (5.15) we know that if the $\vec{\xi}$ variations disappear at times t_0, t_1 than A is extremal only if Euler's equations are satisfied and the boundary term disappears. If on the other hand we make a symmetry displacement i.e. a displacement that makes δA vanish and assume that Euler's equations are satisfied and the boundary term disappears, we obtain that:

$$\int_V \vec{v} \cdot \vec{\xi} d^3\alpha = const. \tag{5.186}$$

this is Noether's theorem in it's fluid mechanical form.

The $\alpha's$ so chosen as to satisfy the equation (5.185) are not unique in fact one can always choose another set of variables say $\tilde{\alpha}'s$ such that:

$$\frac{\partial(\tilde{\alpha}^1, \tilde{\alpha}^2, \tilde{\alpha}^3)}{\partial(\alpha^1, \alpha^2, \alpha^3)} - 1. \tag{5.187}$$

It is quite clear that if the domain of integration is not modified any new set of $\alpha's$ satisfying equation (5.187) can be chosen without effecting the value of the Lagrangian L. This is nothing but Arnold's [24] alpha space symmetry group under which L is invariant (see also Katz & Lynden-Bell [33]). For some flows the domain of integration can be modified without effecting L, in that case we have additional elements in our symmetry group. If we make only small changes $\delta\alpha$ than we can define the group as follows:

$$\frac{\partial \delta\alpha_k}{\partial \alpha_k} = 0, \qquad \delta\vec{\alpha} \cdot \vec{n}\Big|_{surface} = 0, \tag{5.188}$$

where $\vec{n}$ is a unit vector orthogonal to the surface of the alpha space volume which we integrate over. The restriction $\delta\vec{\alpha} \cdot \vec{n}\Big|_{surface} = 0$ is only needed when the infinitesimal transformation changes the domain of integration in such a way as to modify L. In this chapter we are interested in the subgroup of translation, i.e.:

$$\delta\alpha_k = a_k, \qquad a_k = const. \tag{5.189}$$

This subgroup of course does not satisfy $\delta\vec{\alpha} \cdot \vec{n}\Big|_{surface} = 0$ unless at least few of the $\alpha's$ are cyclic or L is not effected by the modification of domain.

In section 1.2.3 we have defined the three following parameters: the magnetic load λ, the magnetic metage μ and η. Notice that since the magnetic lines are closed μ is an angular variable and we can translate it with out changing L. Choosing $\alpha^k = \chi, \eta, \mu$, and inserting those variables into equation (5.185) we re-derive equation (5.107).

The appropriate $\vec{\xi}$ symmetry displacement associated with the infinitesimal change in α_k is given by equation (5.111). For a metage displacement $\vec{\xi}$ takes the form:

$$\vec{\xi} = -\frac{\partial \vec{r}}{\partial \mu}\delta\mu = -\delta\mu\frac{\vec{B}}{\rho}. \tag{5.190}$$

Inserting this expression into the boundary term in equation (5.15) will result in:

$$\delta A_B = \int dt \oint d\vec{S} \cdot [\vec{B}(\frac{1}{2}\vec{v}^2 - w(\rho)) - \vec{v}(\vec{v} \cdot \vec{B})] = 0, \tag{5.191}$$

which is indeed Moffat's condition for magnetic cross helicity conservation [28] as expected. Inserting equation (5.190) into equation (5.186) we obtain the conservation law:

$$\int_V \vec{v} \cdot \frac{\partial \vec{r}}{\partial \mu} d^3\alpha = \int_V \vec{v} \cdot \vec{B} d^3x = \mathcal{H}_C. \tag{5.192}$$

Thus we conclude that the alpha translation group in the direction of μ generates conservation of helicity. [One could of course introduce the symmetry displacement $\vec{\xi} = \epsilon \frac{\vec{B}}{\rho}$, however, in this case one should show that the above displacement is a symmetry group displacement which is not obvious if we do not take into account Arnold's group and Lynden-Bell & Katz labelling. Moreover in coordinate space the symmetry group appears arbitrary complex depending on the flow considered as opposed to its apparent simplicity in alpha space.]

5.8 Conclusion

In this chapter we have reviewed variational principles for barotropic magnetohydrodynamics given by previous authors both in Lagrangian and Eulerian form. Furthermore, we introduced our own Eulerian variational principles from which all the relevant equations of barotropic magnetohydrodynamics can be derived and which are in some sense simpler than those considered earlier. The variational principle was given in terms of four independent functions for non-stationary flows and three independent functions for stationary flows. This is less then the seven variables which appear in the standard equations of magnetohydrodynamics which are the magnetic field $\vec{B}$ the velocity field $\vec{v}$ and the density ρ. In accordance with the theme of this book variational analysis enables the reduction of the number of quantities and equations needed to describe a given physical phenomena, in this chapter this phenomena is magnetohydrodynamics.

The equations in the six function non-stationary case appear have some resemblance to the equations deduced in a previous paper by Frenkel, Levich and Stilman [16]. However, in this previous work the equations were deduced from a postulated Hamiltonian. In the current work we show how this Hamiltonian can be obtained from our simplified Lagrangian using the canonical Hamiltonian formalism.

The appearance of a non zero magnetic helicity and cross helicity, is connected with the fact that some of the functions which we defined are non-single valued. This was elaborated to some extent in the final section of this chapter and was connected to the properties of the functions ζ, ν. We have also shown that the density of cross helicity per unit of magnetic flux is also a conserved quantity and is equal to the discontinuity of ν. Furthermore, we have shown that the conservation of cross helicity can be deduced using the Noether theorem from the symmetry group of magnetic metage translations.

Anticipated applications including new analytic solutions, stability analysis and the description of numerical schemes using the described variational principles, exceed the scope of this chapter.

Some analytical results for magnetic field on a torus were described, however, a full analytic solution in terms of the present formalism for a tokamak as described by Boozer [34] is yet to be constructed.

It was shown by the author that variational principles can be used directly for numerical analysis (simulation) without the need to refer to the field equations. This mathematical construction may lead to better algorithms for simulating magnetohydrodynamics in terms of the needed computer memory and CPU time. This approach was applied to potential flows in a series of papers [35, 36, 37]. A variational formalism of magnetohydrodynamics should serve the same use.

As for stability analysis I suspect that for achieving this we will need to add additional constants of motion constraints to the action as was done by [38], hopefully this will be discussed in a future paper.

Bibliography

[1] C. Eckart, "Variation Principles of Hydrodynamics," *Phys. Fluids*, vol. 3, 421, 1960.

[2] W.A. Newcomb , "Lagrangian and Hamiltonian Methods in Magnetohydrodynamics," *Nucl. Fusion: Supplement*, pt 2, 451-463, 1962.

[3] P.A. Sturrock , Plasma Physics. Cambridge University Press: Cambridge, 1994.

[4] B. Davydov, "Variational principle and canonical equations for an ideal fluid," *Doklady Akad. Nauk*, vol. 69, 165-168, 1949. (in Russian)

[5] V.E. Zakharov, and E.A. Kuznetsov , "Variational Principle and Canonical Variables in Magnetohydrodynamics," *Doklady Akad. Nauk*, vol. 194, no. 6, 1288-1289, 1970. (in Russian). The paper was translated to English: V.E. Zakharov, and E.A. Kuznetsov , "Variational Principle and Canonical Variables in Magnetohydrodynamics," *Soviet Physics - Doklady*, vol. 15, no. 10, 913-914, 1971.

[6] M. Calkin, "An Action Prinicple for Magnetohydrodynamics," *Canadian J. Phys.*, vol. 41, 2241-2251, 1963.

[7] V.A. Vladimirov, and H.K. Moffatt, "On general transformations and variational principles for the magnetohydrodynamics of ideal fluids. Part 1. fundamental principles," *J. Fluid. Mech.*, vol. 283, 125-139, 1995.

[8] A.V. Kats , "Variational principle in canonical variables, Weber transformation, and complete set of the local integrals of motion for dissipation-free magnetohydrodynamics," *JETP Lett*, vol. 77, 657, 2003. Available: Los Alamos Archives physics 0212023.

[9] A.V. Kats, and V.M. Kontorovich, "Hamiltonian description of the motion of discontinuity surfaces," *Low Temp. Phys.*, vol. 23, 89, 1997.

[10] A.V. Kats, "Variational principle and canonical variables in hydrodynamics with discontinuities," *Physica D*, vol. 152-153, 459-474, 2001.

[11] A.V. Kats "Variational principles and canonical variables for MHD flows with discontinuities. I," *Radiofizika and Radioastronomia*, vol. 7, 232, 2002.

[12] T. Sakurai, "A New Approach to the Force-Free Field and Its Application to the Magnetic Field of Solar Active Regions," *Pub. Ast. Soc. Japan*, vol. 31, 209, 1979.

[13] W.H. Yang, P.A. Sturrock, and S. Antiochos, "Force-free magnetic fields - The magneto-frictional method," *Ap. J.*, vol. 309, 383, 1986.

[14] A. Yahalom, and D. Lynden-Bell, "Simplified Variational Principles for Barotropic Magnetohydrodynamics," *Journal of Fluid Mechanics*, vol. 607, 235-265, 2008. Available: Los-Alamos Archives - physics/0603128.

[15] A. Yahalom , "Using Magnetic Fields to Optimize Material Flow - a Variational Approach to Magnetohydrodynamics," In: Proceedings of the Israeli - Russian Bi-National Workshop, Jerusalem: Israel, June 2007.

[16] A. Frenkel, E. Levich, and L. Stilman, "Hamiltonian description of ideal MHD revealing new invariants of motion," *Phys. Lett. A*, vol. 88, 461, 1982.

[17] V.E. Zakharov, and E.A. Kuznetsov, "Hamiltonian Formalism for Nonlinear Waves," *Usp. Fiz. Nauk*, vol. 40, 1087, 1997.

[18] P. Morrison , "Poisson Brackets for Fluids and Plasmas," *American Inst. Phys. Conf. Proc.*, vol. 88, 13, 1982.

[19] M.D. Kruskal, and Kulsrud R M., "Equilibrium of Magnetically Confined Plasma in a Toroid," *The Physics of Fluids*, vol. 1, no. 4, 265, 1958.

[20] A. Yahalom "A Four Function Variational Principle for Barotropic Magnetohydrodynamics," *EPL*, 89, 34005, 2010. Available: doi 10.1209/0295-5075/89/34005, Los-Alamos Archives - arXiv:0811.2309.

[21] J.D. Bekenstein, and A. Oron, "Conservation of circulation in magnetohydrodynamics," *Physical Review E*, vol. 62, no. 4, 5594-5602, 2000.

[22] A.V. Kats, "Canonical description of ideal magnetohydrodynamic flows and integrals of motion," *Phys. Rev. E*, vol. 69, 046303, 2004.

[23] V.A. Vladimirov, and H.K. Moffatt, "On general transformations and variational principles for the magnetohydrodynamics of ideal fluids. Part I. Fundamental principles," *J. Fluid Mech.*, vol. 283, 125-139, 1995. V.A. Vladimirov, H.K. Moffatt, and K.I. Ilin, "On general transformations and variational principles for the magnetohydrodynamics of ideal fluids. Part II. Stability criteria for two-dimensional flows," *J. Fluid Mech.*, vol. 329, 187, 1996. V.A. Vladimirov, H.K. Moffatt, and K.I. Ilin, "On general transformations and variational principles for the magnetohydrodynamics of ideal fluids. Part III. Stability Criteria for axisymmetric flows," *J. Plasma Phys.*, vol. 57, 89, 1997. V.A. Vladimirov, H.K. Moffatt, and K.I. Ilin, "On general transformations and variational principles for the magnetohydrodynamics of ideal fluids Part IV. Generalized isovorticity principle for three-dimensional flows," *J. Fluid Mech.*, 390, 127, 1999.

[24] J.A. Almaguer, E. Hameiri, J. Herrera, and D.D. Holm, "Lyapunov Stability Analysis of Magnetohydrodynamic Plasma Equilibria with Axisymmetric Toroidal Flow," *Phys. Fluids*, vol. 31, 1930-1939, 1988.

[25] J.W. Dungey, Cosmic Electrodynamics. Cambridge University Press: Cambridge, 1958.

[26] E.A. Kuznetsov, "Mixed Lagrangian-Eulerian description of vortical flows for ideal and viscous fluids," *Journal of Fluid Mechanics*, vol. 600, 167-180, 2008.

[27] D. Lynden-Bell, "Consequence of one spring researching with Chandrasekahar," *Current Science*, vol. 70, No 9., 789-799, 1996.

[28] H.K. Moffatt, "The degree of knottedness of tangled vortex lines," *J. Fluid Mech.*, vol. 35, 117, 1969.

[29] Moreau, J.J. 1977, *Seminaire D'analyse Convexe, Montpellier 1977 , Expose no: 7*

[30] V.I. Arnold, "Sur un principe variationel pour les ecoulements stationaires," *J. de Mecanique*, vol. 5:1, 29-43, 1966.

[31] A. Yahalom , "Helicity Conservation via the Noether Theorem," *J. Math. Phys.*, vol. 36, 1324-1327, 1995. Available: Los-Alamos Archives solv-int/9407001.

[32] A. Yahalom, "Energy Principles for Barotropic Flows with Applications to Gaseous Disks", PhD Thesis, The Hebrew University of Jerusalem, Israel, 1996.

[33] J. Katz, and D. Lynden-Bell, "The dynamics of isocirculational flows," *Geophysical and Astrophysical Fluid Dynamics*, vol. 33, 1-33, 1985.

[34] A.H. Boozer , "Physics of magnetically confined plasmas," *Rev. Mod. Phys.*, vol. 76, 1071, 2004.

[35] A. Yahalom, and G.A. Pinhasi, "Simulating Fluid Dynamics Using a Variational Principle," In: Proceedings of the AIAA Conference, Reno: USA, 2003.

[36] A. Yahalom , G. A. Pinhasi, and M. Kopylenko, "A Numerical Model Based on Variational Principle for Airfoil and Wing Aerodynamics ," In: Proceedings of the AIAA Conference, Reno: USA, 2005.

[37] D. Ophir, A. Yahalom, G.A. Pinhasi, and M. Kopylenko, "A Combined Variational and Multi-grid Approach for Fluid Simulation," In: Proceedings of International Conference on Adaptive Modelling and Simulation (ADMOS), Barcelona:Spain 2005, pp. 295-304.

[38] A. Yahalom, J. Katz, and S. Inagaki,"Energy Principles for Self-Gravitating Barotropic Flows - Part Two - the Stability of Maclaurin Discs," *Mon. Not. R. Astron. Soc.*, vol. 268, 506-516, 1994.

CHAPTER 6

The geometrical meaning of time - the emergence of the concept of time in the general theory of relativity

Abstract

It is stated in many text books that the any metric appearing in general relativity should be locally Lorentzian i.e. of the type $\eta_{\mu\nu} =$ diag $(1, -1, -1, -1)$ this is usually presented as an independent axiom of the theory, which can not be deduced from other assumptions. In this work we show that the above assertion is a consequence of a standard linear stability analysis of the Einstein equations and need not be assumed.

6.1 Introduction

It is well known that our daily space-time is approximately of Lorentz (Minkowski) type that is, it possesses the metric $\eta_{\mu\nu} =$ diag $(1, -1, -1, -1)$. The above statement is taken as one of the central assumptions of the theory of special relativity.

Further more it is assumed in the general theory of relativity that any space-time is locally of the type $\eta_{\mu\nu} =$ diag $(1, -1, -1, -1)$, although it can not be presented so globally due to the effect of matter. This is a part of the demands dictated by the well known equivalence principle. The above principle is taken to be one of the assumptions of general relativity in addition to the Einstein equations:

$$G_{\mu\nu} = -\frac{8\pi G}{c^4} T_{\mu\nu} \tag{6.1}$$

in which $G_{\mu\nu}$ is the Einstein tensor, $T_{\mu\nu}$ is the stress-energy tensor, G is the gravitational constant and c is the velocity of light.

As described in the preface of this book an advancement in the theory of general relativity would be a reduction in the number of assumptions needed to generate the theory. In what follows we will show that the assumption that space-time locally posses a metric of the type $\eta_{\mu\nu} = \text{diag}\ (1,-1,-1,-1)$ is not necessary, (contrary to what is argued in so many text books, see for example [1]) rather we will argue that this metric is the only possible stable solution to the Einstein equation (6.1) in vacuum, that is for the case $T_{\mu\nu} = 0$. And thus reduce the number of assumptions needed to obtain the celebrated results of general relativity. By making the theory more compact we enhance its predictive strength.

Eddington [2, page 25] has considered the possibility that the universe contains different domains in which some domains are locally Lorentzian and others have some other local metric of the type $\eta_{\mu\nu} = \text{diag}\ (-1,-1,-1,-1)$ or the type $\eta_{\mu\nu} = \text{diag}\ (+1,+1,-1,-1)$. For the first case he concluded that the transition will not be possible since one will have to go through a static universe with a metric $\eta_{\mu\nu} = \text{diag}\ (0,-1,-1,-1)$[1]. Going to the domain in which $\eta_{\mu\nu} = \text{diag}\ (+1,+1,-1,-1)$ means that one will have to pass through $\eta_{\mu\nu} = \text{diag}\ (+1,0,-1,-1)$ in which space becomes two dimensional[2]. The stability of those domains was not discussed by Eddington.

Greensite [3] and Carlini & Greensite [4] have studied the metric $\eta_{\mu\nu} = \text{diag}\ (e^{i\theta},-1,-1,-1)$ in which θ the "wick angle" was treated as a quantum field dynamical variable. They have shown that the real part of the quantum field effective potential is minimized for the Lorentzian metric $\theta = 0$ and for the same case the imaginary part of the quantum field effective potential is stationary. Further more they have calculated the fluctuations around this minimal value and have shown them to be of the order $(\frac{l_p}{R})^3$ in which l_p is the Planck length and R is the scale of the universe. Elizalde & collaborators

[1]Prof. Lynden Bell has noticed that there may be another way going through the metric $\eta_{\mu\nu} = \text{diag}\ (\infty,-1,-1,-1)$, the author thanks him for his remark.

[2]Again there may be another way going through the metric $\eta_{\mu\nu} = \text{diag}\ (+1,\infty,-1,-1)$.

[5] have shown that the same arguments apply to a five dimensional Kaluza-Klein universe of the type $R^4 \times T^1$.

Itin & Hehl [6] have deduced that space time must have a Lorentzian metric in order to support classical electric/magnetic reciprocity.

H. van Dam & Y. Jack Ng [7] have argued that in the absence of a Lorentzian metric one can not obtain an appropriate finite representation of the relevant groups and hence the various quantum wave equations can not be written.

What is common to the above approaches is that additional theoretical structures & assumptions are needed in order to justify what appears to be a fundamental property of space-time. In this chapter we claim otherwise. We will show that General relativistic equations and some "old fashioned" stability analysis will lead to a unique choice of the Lorentzian metric being the only one which is stable.

The plan of this chapter is as follows: in the first section we describe the possible constant metrics which are not equivalent to each other by trivial manipulations. The second section will be devoted to the review the classical linearized theory of gravity which we adapt to our more general case of a non Lorentzian background metric. In the third section we use the linearized theory to study the stability of the possible constant metrics and thus divide them into two classes: stable and unstable. The last section will discuss some possible implications of our results.

6.2 Possible Constant Metrics

In this section we study what are the possible constant metrics available in the general theory of relativity which are not equivalent to one another by a trivial transformation, that amounts to a simple change of coordinates.

Let us thus study the four-dimensional interval:

$$d\tau^2 = \eta_{\mu\nu} dx^\mu dx^\nu \tag{6.2}$$

in what follows Greek letters take the traditional values of $1-4$, and summation convention is assumed. $\eta_{\mu\nu}$ is any real constant matrix.

Since $\eta_{\mu\nu}$ is symmetric we can diagonalize it using a unitary transformation in which both the transformation matrix and the eigenvalues obtained are real. Thus without loss of generality we can assume that in a proper coordinate basis:

$$\eta = \text{diag}\ (\lambda_1, \lambda_2, \lambda_3, \lambda_4). \tag{6.3}$$

Next, by changing the units of the coordinates, we can always obtain:

$$\eta = \text{diag}\ (\pm 1, \pm 1, \pm 1, \pm 1) \tag{6.4}$$

notice that a zero eigen-value is not possible due to our assumption that the space is four dimensional.

We conclude that the metrics η given in equation (6.4) are the most general constant metrics possible. In what follows we will study the stability of those solutions.

6.3 Stability of Constant Metrics

To study the stability of the metric $\eta_{\mu\nu}$ given in equation (6.4) we make an arbitrary small perturbation of the constant metric and obtain the perturbed metric $g_{\mu\nu}$.

$$g_{\mu\nu} = \eta_{\mu\nu} + h_{\mu\nu}. \tag{6.5}$$

The evolution of the metric g is studied under the Einstein equation (6.1) . In what follows we adapt the well known linearized theory of gravity given in many text books such as [1], to our more generalized needs. In order to obtain the linearized equations the following quantities should be linearized:

1. The connection coefficients deduced from the metric:

$$\Gamma^{\mu}_{\alpha\beta} \equiv \frac{1}{2} g^{\mu\nu}(g_{\alpha\nu,\beta} + g_{\beta\nu,\alpha} - g_{\alpha\beta,\nu}) \tag{6.6}$$

   in which , stand for partial derivative, and $g^{\mu\nu}$ is the inverse matrix of $g_{\mu\nu}$.

2. The Ricci tensor $R_{\mu\nu}$ which is deduced from the connection coefficients:

$$R_{\mu\nu} \equiv \Gamma^{\alpha}_{\mu\nu,\alpha} - \Gamma^{\alpha}_{\mu\alpha,\nu} + \Gamma^{\alpha}_{\beta\alpha}\Gamma^{\beta}_{\mu\nu} - \Gamma^{\alpha}_{\beta\nu}\Gamma^{\beta}_{\mu\alpha} \tag{6.7}$$

3. The Einstein tensor $G_{\mu\nu}$ which is deduced from the Ricci tensor and the Curvature Scalar:

$$G_{\mu\nu} \equiv R_{\mu\nu} - \frac{1}{2} g_{\mu\nu} R, \qquad R \equiv g^{\mu\nu} R_{\mu\nu} \tag{6.8}$$

6.3.1 Notations

First we introduce some notations: the inverse metric of $g_{\mu\nu}$ is $g^{\mu\nu}$ given by:

$$g^{\mu\nu} = \eta^{\mu\nu} - h^{\mu\nu} \tag{6.9}$$

in which $\eta^{\mu\nu}$ is the inverse matrix of $\eta_{\mu\nu}$, by virtue of equation (6.4) it is also identical to it i.e., $\eta^{\mu\nu} = \eta_{\mu\nu}$. An easy calculation will show that to first order in h we obtain:

$$h^{\mu\nu} = \eta^{\sigma\nu}\eta^{\rho\mu}h_{\sigma\rho}. \tag{6.10}$$

Further more we introduce the following notations:

$$h^{\mu}_{\sigma} = \eta^{\mu\nu}h_{\sigma\nu} \qquad h = h^{\mu}_{\mu} = \eta^{\mu\nu}h_{\mu\nu}. \tag{6.11}$$

Generally speaking we use the constant metric η to raise and lower indices.

6.3.2 The Connection Coefficients

Let us now calculate the linearized form of the affine connection which is given by equation (6.6) . Inserting equation (6.5) and equation (6.9) and keeping only the first order terms in h we obtain:

$$\Gamma^{\mu}_{\alpha\beta} = \frac{1}{2}\eta^{\mu\nu}(h_{\alpha\nu,\beta} + h_{\beta\nu,\alpha} - h_{\alpha\beta,\nu}) \tag{6.12}$$

which can also be written using equation (6.11) as:

$$\Gamma^{\mu}_{\alpha\beta} = \frac{1}{2}(h_{\alpha}{}^{\mu}{}_{,\beta} + h_{\beta}{}^{\mu}{}_{,\alpha} - h_{\alpha\beta,}{}^{\mu}) \tag{6.13}$$

6.3.3 The Ricci Tensor

The Ricci Tensor given in equation (6.7) can be written in a linearized form using the result of equation (6.13) and the notation defined in equation (6.11)

$$R_{\mu\nu} = \frac{1}{2}(h_{\mu}{}^{\alpha}{}_{,\alpha\nu} + h_{\nu}{}^{\alpha}{}_{,\alpha\mu} - h_{\mu\nu,\alpha}{}^{\alpha} - h_{,\mu\nu}). \tag{6.14}$$

The linearized Curvature Tensor can be calculated from equation (6.14)

$$R = \eta^{\mu\nu}R_{\mu\nu} = h^{\mu\alpha}{}_{,\mu\alpha} - h_{,\alpha}{}^{\alpha}. \tag{6.15}$$

6.3.4 The Einstein Tensor

Finally we obtain the linearized form of the Einstein Tensor which can be calculated from equations (6.8,6.14,6.15):

$$G_{\mu\nu} = \frac{1}{2}(h_{\mu}{}^{\alpha}{}_{,\alpha\nu} + h_{\nu}{}^{\alpha}{}_{,\alpha\mu} - h_{\mu\nu,\alpha}{}^{\alpha} - h_{,\mu\nu}) - \frac{1}{2}\eta_{\mu\nu}(h_{\alpha\beta,}{}^{\alpha\beta} - h_{,\alpha}{}^{\alpha}). \quad (6.16)$$

In order to simplify the above notation the following quantity is defined:

$$\bar{h}_{\mu\nu} = h_{\mu\nu} - \frac{1}{2}\eta_{\mu\nu}h \quad (6.17)$$

Using this definition, equation (6.16) can be written as:

$$2G_{\mu\nu} = -\bar{h}_{\mu\nu,\alpha}{}^{\alpha} - \eta_{\mu\nu}\bar{h}_{\alpha\beta,}{}^{\alpha\beta} + \bar{h}_{\mu\alpha,}{}^{\alpha}{}_{\nu} + \bar{h}_{\nu\alpha,}{}^{\alpha}{}_{\mu}. \quad (6.18)$$

6.3.5 Gauge Transformation

The metric being a tensor can always be transformed to another coordinate system by the transformation:

$$g'_{\sigma\rho}(x') = g_{\mu\nu}\frac{\partial x^{\mu}}{\partial x'^{\sigma}}\frac{\partial x^{\nu}}{\partial x'^{\rho}} \quad (6.19)$$

Let us introduce the transformation:

$$x^{\mu} = x'^{\mu} - \xi^{\mu} \quad (6.20)$$

in which ξ is same order of magnitude as h. Further more let us define:

$$g'_{\mu\nu}(x') = \eta_{\mu\nu} + h'_{\mu\nu}(x). \quad (6.21)$$

Inserting equations (6.5,6.21,6.20) into equation (6.19) we obtain to first order in h:

$$h'_{\mu\nu} = h_{\mu\nu} - \xi_{\mu,\nu} - \xi_{\nu,\mu} \qquad (\xi_{\mu} = \eta^{\mu\nu}\xi_{\nu}) \quad (6.22)$$

this is denoted as the "gauge transformation", by changing the coordinates infinitesimally we can always obtain a new h' which is different from the old h and is related to it by equation (6.22) . Our obvious choice of gauge will be one that simplifies equation (6.18) . Let us calculate the expression $\bar{h}'_{\mu\alpha,}{}^{\alpha}$:

$$\bar{h}'_{\mu\alpha,}{}^{\alpha} = \bar{h}_{\mu\alpha,}{}^{\alpha} - \xi_{\mu,\alpha}{}^{\alpha}. \quad (6.23)$$

We can always choose ξ such that:

$$\xi_{\mu,\alpha}{}^{\alpha} = \bar{h}_{\mu\alpha,}{}^{\alpha} \tag{6.24}$$

in which the equation above is a second order equation for the ξ_μ's. This does not "fix" the gauge since we can always introduce a new gauge: $\xi^1_\mu = \xi_\mu + \xi^0_\mu$ in which: $\xi^0_{\mu,\alpha}{}^{\alpha} = 0$. Choosing the gauge according to equation (6.24) we obtain:

$$\bar{h}'_{\mu\alpha,}{}^{\alpha} = 0 \tag{6.25}$$

writing equation (6.18) in terms of h' and dropping the prime we see that:

$$2G_{\mu\nu} = -\bar{h}_{\mu\nu,\alpha}{}^{\alpha}. \tag{6.26}$$

6.4 Stability Analysis

In the lack of matter Einstein equation (6.1) becomes $G_{\mu\nu} = 0$ that is through equation (6.26) we obtain the following equations for $\bar{h}_{\mu\nu}$:

$$\bar{h}_{\mu\nu,\alpha}{}^{\alpha} = 0. \tag{6.27}$$

Next we introduce the Fourier decomposition of $\bar{h}_{\mu\nu}$:

$$\bar{h}_{\mu\nu} = \frac{1}{(2\pi)^{\frac{3}{2}}} \int_{-\infty}^{\infty} A_{\mu\nu}(x_0, \vec{k}) e^{i\vec{k}\cdot\vec{x}} d^3k, \quad \vec{k} = (k^1, k^2, k^3), \quad \vec{x} = (x^1, x^2, x^3) \tag{6.28}$$

Introducing the decomposition equation (6.28) into equation (6.27) leads to:

$$\eta^{00}\partial_0^2 A_{\mu\nu} - \eta^{ij} k_i k_j A_{\mu\nu} = 0 \tag{6.29}$$

in which i, j are integers between $1-3$. Choosing $\eta^{00} = 1$ we see that the only way to avoid exploding solutions is to choose $\eta^{ij} = \text{diag}\ (-1,-1,-1)$, thus one stable metric would be:

$$\eta^{(1)} = \text{diag}\ (1,-1,-1,-1) \tag{6.30}$$

alternatively we can choose $\eta^{00} = -1$ in this case the only way to avoid exploding solutions is to choose $\eta^{ij} = \text{diag}\ (1,1,1)$, thus a second stable metric would be:

$$\eta^{(2)} = \text{diag}\ (-1,1,1,1) \tag{6.31}$$

that is $\eta^{(1)} = -\eta^{(2)}$.

A famous example due to Hadamard shows that exploding solutions appear when the Cauchy conditions are applied to the Laplace equation. This example is in fact the reason behind part of the author's results. When one can not prevent nature from enforcing arbitrary Cauchy type conditions one is forced to give up Euclidean (and other types) of flat metrics to avoid Laplace like equations and hence exploding solutions.

One can claim that even in the Lorentzian case we do have an exploding solution $A_{\mu\nu} \sim x_0$ in the degenerate case $|\vec{k}| = 0$ in which the perturbation is uniform over the entire manifold $x_0 = constant$. Notice, however, this solution is only linearly exploding and not exponentially exploding as is the generic case of the other flat metrics considered. A way to avoid this type of solution is to assume that space time is topologically a 4D Torus. In this case, this type of solution may be rejected on the grounds that the metric must be a single valued function of the space time coordinates.

The reader should notice that only the "spatial" functions A_{mn} should be thought of as independent. The "temporal" function $A_{0\mu}$ are dependent on A_{mn} through equations (6.25), which takes the form:

$$\partial_0 A_{0n} + i\frac{\eta^{rs}}{\eta^{00}} k_r A_{ns} = 0, \qquad \partial_0 A_{00} + i\frac{\eta^{rs}}{\eta^{00}} k_r A_{0s} = 0 \tag{6.32}$$

It is easy to show that equation (6.32) can be solved simultaneously with equation (6.29) for the components $A_{0\mu}$.

In the case that the universe has a spatial cyclic topology in one or more directions the Fourier integral in this direction can be replaced by a Fourier series such that we only have k_i's of the type:

$$k_i = \frac{2\pi n_i}{L_i} \tag{6.33}$$

in which n_i is an integer and L_i is the dimension of the spatially cyclic universe in the i direction.

Notice that when $\bar{h}_{\mu\nu}$ becomes large, the linearized Einstein equations are no longer appropriate. It is then necessary to study the full, nonlinear Einstein equations to determine if the solution in question continues diverging. It is possible to imagine that a solution that appears to diverge is actually well-behaved when studied in full. Although linear theory seems most appropriate for most astronomical phenomena excluding extreme cases such as black holes one can not neglect the mathematical possibility that non-linear terms are important. However, this chapter deals with linear stability and the study of non-linear stability is left for future endeavors probably using numerical methods.

It must be stressed that "linear stability" is not an assumption. Solutions that are not linearly stable must develop under the equations due to the slightest perturbation to something else. Thus they cannot be observed in reality. The formal meaning of linear stability which is ubiquitous in the literature is as follows. Linear stability means that a small perturbation will not grow under the linear equations and will therefore remain bound and small. An instability means that the equations will cause an initially small perturbation to become unbound and eventually not small. An instability can be either exponentially fast growing as is the case in non-Lorentzian metrics or linearly slow growing as in the case of a Lorentzian metric in either case it will not remain small. The question is, since the Lorentzian metric is not stable why is it still here. The answer is two fold: one possibility is that it will not stay here for long but the instability is slow so we may observe it for a while, the second possibility is more intriguing since it contains information not just on the local geometry of space time but on the topology of space time as well. The question of the possible topology of space-time is not settled at this time. And the current analysis leads to a new angle regarding this problem. On a topological torus the Lorentzian metric will remain constant for ever.

Notice that the instabilities discussed in this section are not mere coordinate instabilities as they lead to growth in gravitational waves and components of the Riemann tensor.

6.5 Conclusions

The historical development of the general theory of relativity can be succinctly descried as follows. First, Maxwell established his set of equations based on well established empirical facts. Then it was noticed that the equations do not respect the Galileo group of transformations but rather the Lorentz group of transformations. The new physics, consisting of Maxwell's electrodynamics and Einstein's mechanics, then satisfied a new principle of relativity, the Principle of Special Relativity, which says that all physical equations must be invariant under Lorentz transformations [9]. The development of General relativity followed later when Einstein incorporated gravity into the relativity frame-work, still demanding that locally the space-time metric should be of Lorentz type.

From a fundamental point of view the derivation should be somewhat different. That is, one may imagine that space-time is a geometrical manifold and ask how should such a manifold arrange itself. The behavior of such a manifold could be contemplated to emerge from a variational principle in which the action is

$$\int \sqrt{g} R d^4 x, \tag{6.34}$$

in which g is the metric Jacobean and R is the curvature scalar (the manifold will try to be as flat as possible). This variational principle will yield the Einstein vacuum field equations. Having derived the equations one may ask what are the stable solutions of such equations. Taking into account that a manifold must have a tangent space and hence locally must have a constant metric, one may ask what type of constant metrics can exist. The arguments outlined in the chapter suggests that such a metric should be Lorentz. Only after specifying the properties of an empty space-time one is allowed to contemplate what type of entities such as matter and fields can inhabit such a manifold, leading perhaps to Maxwell's electrodynamics and Einstein's mechanics. Thus the fundamental derivation takes an opposite route than the historical one.

We conclude from equations (6.30,6.31) that the only constant stable solution is of a Lorentz (Minkowski) type.

For other constant solutions we expect instabilities for $k_i \to \infty$ where i depends on the unstable solution chosen. Thus the instabilities vary on very small length scale of which $\lambda = \frac{2\pi}{k} \to 0$, this length can be the smallest for which the general theory of relativity is applicable, perhaps the planck scale $\lambda = l_p = 1.61610^{-35} m$, in that case an unstable solution will last for about $t = \frac{\lambda}{c} = 5.3910^{-44} sec$. However, in the presence of matter this may take longer. This may explain why in QED an unstable Euclidean metric is used such that $\eta = \text{diag}\,(1,1,1,1)$, this is referred to as "Wick's rotation" [8].

This work gives a plausible explanation why the flat metric of the 4 dimensional space of general relativity is Lorentzian. And hence explains the existence of the intuitive partition of this 4 dimensional space into "spatial" space and "temporal" time. The existence of a such a partition in an almost empty space-time does not contradict the fact that such a partition can not be demonstrated in general solutions of equation (6.1) such as the one discovered by Gödel [10]. The main difference between the solutions studied in this chapter and the solutions studied by Gödel, is that in this chapter we study solutions of a vacuum theory while in Gödel's analysis a distribution of matter is assumed. What is claimed is that the concept of time is meaningful in moderate mass distributions scenarios in which the gravitational field is weak which is the prevalent case in astrophysical scenarios. However, there is no guarantee the one can meaningfully discuss time under extreme gravitational fields (and extreme mass distributions) as in the cases discussed by Gödel.

The reader may inquire why can't 3D theories such as geometrostatics and electrostatics be subject to the same analysis presented in this chapter? The reason is the following: Geometrostatics and electrostatics are differential presentations of Newton's inverse square law and Coulomb's inverse square law and hence they are empirical in nature. Those theories contain in addition to Poisson's equation also the boundary conditions of this equation at infinity (that is the fields should vanish far away from the source). Under those conditions it is well known that well behaved solutions exist.

On the other hand the equations of the general theory of relativity which is considered to be a fundamental theory derived from first principles should also have "boundary conditions" derived from first principles. Since the theory is local in nature it is only reasonable to assume that the "boundary conditions" should be also local and not "global" dictating a certain behavior at "infinity". This of course does not dictate what type of values the linear metric should take locally and thus we have a freedom to choose arbitrary values which lead to stability analysis described in this chapter and to the result that only the Lorentz metric can be considered stable.

Bibliography

[1] C.W. Misner, K.S. Thorne, and J.A. Wheeler, Gravitation. W.H. Freeman and Company Limited, Reading, England, 1973.

[2] A.S. Eddington, The mathematical theory of relativity. Cambridge University Press: Cambridge, 1923.

[3] J. Greensite, "Dynamical Origin of the Lorentzian Signature of Spacetime," *Phys. Lett. B*, vol. 300, 34-37, 1993. Available: Los Alamos Archive gr-qc/9210008.

[4] A. Carlini, and J. Greensite, "Why is Spacetime Lorentzian?," *Phys. Rev. D*, vol. 49, 866-878, 1994. Available: Los Alamos Archive gr-qc/9308012 (12 Aug 1993).

[5] E. Elizalde, S.D. Odintsov, and A. Romeo, "Dynamical determination of the metric signature in spacetime of non-trivial topology," *Class. Quantum Grav.*, vol. 11, no. 4, L61-L67, 1994.

[6] Y. Itin, and F.W. Hehl, "Is the Lorentz signature of the metric of spacetime electromagnetic in origin?," *Annals Phys.*, vol. 312, 60-83, 2004). Available: Los Alamos Archive gr-qc/0401016 (6 Jan 2004).

[7] H. van Dam, and Y. Jack Ng, "Why 3 + 1 metric rather than 4 + 0 or 2 + 2?," *Phys. Lett. B*, vol. 520, 159-162, 2001). Available: Los Alamos Archive hep-th/0108067 (10 Aug 2001).

[8] S. Weinberg, The Quantum Theory of Fields. Cambridge University Press: Cambridge, 1995.

[9] S. Weinberg, Gravitation and Cosmology: Principles and Applications of the General Theory of Relativity. John Wiley and Sons, Inc.: New York, 1972.

[10] P. Yourgrau, A World Without Time. Basic Books: New York, 2006.

CHAPTER 7

Modified Newtonian Dynamics (MOND) in the framework of covariant and non covariant field theory

Marcelo Schiffer[1]

Abstract

"In a spiral galaxy, the ratio of dark-to-light matter is about a factor of ten. That's probably a good number for the ratio of our ignorance-to-knowledge. We're out of kindergarten, but only in about third grade."
Vera Rubin

7.1 Introduction

The general theory of relativity (GR) is known to reproduce a significant amount of useful results which were verified by observation. Nevertheless, some recent (and less recent) observations can not be understood in terms of GR and observed matter. Thus, an advancement in the field theory of

[1]Ariel University Center of Samaria, P.O.B. 3, Ariel 40700 Israel, e-mail: schiffer@ariel.ac.il

Asher Yahalom (Ed)

gravity would be a generalization of GR which will take into account all observations. Such generalizations, their advantages and limitations are to be discussed below.

7.2 The Advent of Dark Matter

The advent of General Relativity resolved the anomalous precession of the orbit of Mercury and predicted the bending of light by the sun, as confirmed by Eddington in 1916. General Relativity replaced Newtonian paradigm which was held for 232 years.

As General Relativity was able to predict correctly:

- Solar System Observations: precessions and light bending...
- Lensing Effect: Bending of light from distant objects by intervening galaxies
- Cosmological observations: Expansion of the Universe, abundance of light elements,the background radiation, ...

it became the new paradigm for gravitation. Nevertheless, first fissures emerged from cosmological and astrophysical observations. We number a few, at different scales:

- In 1933 Fritz Zwicky realized that the velocities of the Galaxies within the Comma Cluster are way larger than those predicted by the virial theorem in Newtonian theory[1]. He remarked that the amount of matter needed to account for the velocities could be 400 times that of the visible matter. He postulated the existence of an unseen form of matter permeating the cluster. Volders in 1959 remarked that stars in the periphery of the neighbor spiral galaxy M33 do not move as expected [2]. The virial theorem in Newtonian Gravity predicts that $MG/r \sim Mv^2$, that is to say, the rotation curve should increase and at some point bend down, and the velocity should drop off as $1/\sqrt{r}$. In the seventies Rubin and Ford [3],[4] showed for a very large sample of spiral galaxies that this behavior is a general feature: velocities at the periphery of the galaxies do not bend down , attain a plateau at some velocity for each galaxy. In the figure below (figure 7.1) we see a bunch of rotation curves for some galaxies.

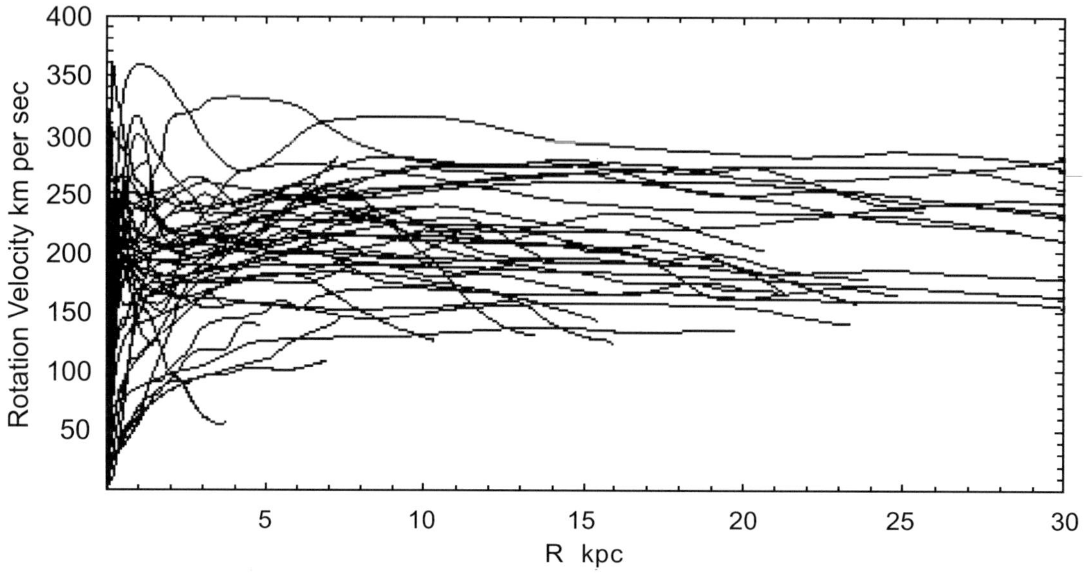

Figure 7.1: From Sofue Y. and Rubin V, Annu. Rev. Astron. Astrophys. 2001. 39: 137-174.

Deviation from Newtonian behavior occurs at a nearly fixed acceleration scale $a_0 \approx 2 \times 10^{-8} cm/s^2$. In 1977, Tully and Fisher [5] put forward their empirical law that the Luminosity of the galaxy scales with the fourth power of the plateau velocity. Clearly, this behavior cannot be explained by GR or Newtonian theory, its nonrelativistic limit, from the observable mass in the galaxy. Should a dark matter content exist, according to Poisson's equation its density should fall off as $\rho_{dark} \sim 1/r^2$. Measurements of the velocity dispersions in elliptical galaxies also confirmed the existence of dark matter [6]

- According to GR radiation coming from quasar or bright galaxy should be bent by intervening galaxies or cluster of galaxies. The image should look like either multiple images or as an arc, depending upon the geometrical configuration (strong lensing). The angular separation of the images is an indication of the amount of intervening mass The observable matter among dozens of clusters observed cannot account for the amount of lensing, again dark matter is called for. The amount of dark matter that reproduces the lensing agrees with that needed to explain velocity dispersions within each one of the clusters (again, inferred from the virial's theorem) [7].

Perhaps, the most convincing evidence of dark matter comes from collision of clusters of galaxies. As they cross, the hot ionized gases in the galaxies interact strongly, producing X-ray emissions. The other components of the galaxies, dark matter and stars should pass each other nearly undisturbed. They overshoot and then are pulled back by gravitational attraction. Should we exclude dark matter, 90% of the cluster is to be found in the hot gas. Thus, the mapping of the mass distribution of the cluster via gravitational lensing should be centered around the hot X-ray emitting gas. Nevertheless, gravitational lensing produced by the "bullet" cluster (1E 0657-56) reveals two lensing centers away from the visible mass, exactly at the places where one would expect dark matter components would reside [8],[9]. Further support for dark matter follows MACS J0025.4-1222 [10] which is the remnant from the collisions of the cluster of galaxies. Again the lensing indicates the presence of dark matter.

- The flatness problem, the monopole abundance predicted by grand unified theories and the homogeneity of radiation among regions which where never in causal contact could not be explained by the standard cosmological model. In 1980 Guth [11] introduced the inflationary hypothesis, which required either a cosmological constant or an unseen form of energy, the dark energy, which can be modelled by symmetry breaking mechanisms in particle physics. With the refinement of the measurements of the anisotropy of background radiation, and the use of type Ia supernovae as standard candles it was possible to extract crucial information from the peaks of the multipole distribution of the correlation functions of the background radiation [12] which together with the fading of light coming from the aforementioned supernovae revealed us that we live in an accelerating Universe filled with 75% of dark energy, 24% of dark matter and the remaining 4% of barionic matter [13].

7.3 The first dark matter paradigm

At the end of the 18th century Le Verrier faced the challenge of predicting to a very large precision the orbits of the planets in the solar system.

When faced with Mercury 565 seconds precession of the orbit, he calculated the effect of the outer planets which accounted for 527 seconds.As he could not account for the remaining 38 seconds, he moved to the problem of the observed perturbations of the orbit of Uranus. In 1846 he showed that a putative planet with given mass and orbit would account for the perturbations. Astronomers pointed their telescope to the place predicted and found Neptune exactly where it was supposed to be, only three weeks later. The dark matter paradigm was born. Newtonian gravity was at its peak, and Le Verrier returned to Mercury's problem convinced that a inter mercurial planet exists. He postulated Vulcan, which was never observed. His belief on the accuracy of Newton's Law of Gravity was so solid that he died convinced in the existence of a "dark matter" in the vicinity of the Sun. The rest of the story is well known. Deciding towards a "dark matter" was then relatively easy, the putative object had either to be observed, or alternatively, through other effects it produced.

So far, the unique evidence for dark matter is its gravitational pull and its composition is elusive . Furthermore, since one fills up space with dark matter with the distribution and the properties needed to reproduce the gravitational discrepancies, its seems hopeless from epistemological point of view to falsify the dark matter paradigm.

That might not be the case:

- Two thirds of all disk galaxies host a barlike structure formed by stars. The baris formed via gravitational instability: either Numerical N-body simulations or density perturbation show that in the presence of disks the bar-structure necessarily forms. A massive halo (dark matter) would make the disk geometry irrelevant and practically suppress the yet observed formation of bars [14].

- Numeric simulations of dark matter distribution predict a $1/r$ behavior near the bulge and a $1/r^2$ at the outskirts of the galaxy. Tailoring the distribution of dark matter to reproduce the flat rotation curves (tuning halo parameters) is also problematic. Furthermore, it is hard to explain Tully Fisher's unless one again fine tunes dark matter parameters.

  Is that conceivable that we might have to face a rupture of the General Relativity paradigm?

7.4 Mond

In 1983 Milgrom proposed an alternative to dark matter, MOND. It relies upon a modification of the inertial law [15]:

$$\mu\left(|\vec{a}/a_o|\right)\vec{a} = -\vec{\nabla}\Phi_N \tag{7.1}$$

where μ is some function of $\vec{a}$, the particle's acceleration, Φ_N is the Newtonian potential obtained from Poisson's equation:

$$\nabla^2\Phi_N = 4\pi G\rho \tag{7.2}$$

where ρ is the density of barionic matter. It is postulated that $\mu(x) \rightarrow 1$ as $x >> 1$, in which way we recover Newtonian Dynamics, and $\mu(x) = \alpha x$ for $x << 1$ (α is a constant of order 1). With this choice and far from the bulk of mass, in a region where $\Phi_N = -GM/r$, $\alpha(v^2/r)^2/a_0 = GM/r^2$ we obtain a constant velocity profile with velocity $v_0^4 = a_0 GM/\alpha$. Thus , MOND not only predicts the plateau in the rotation curves but the also Tully-Fisher's law, if we assume that light tracks matter. The proportionality coefficient $GMa_0 \approx 50M_\odot(km/s)^4$ agrees with the measured coefficient in Tully Fisher's law. In summary, assuming MOND's modification of the law of inertia, a single parameter theory, we can explain:

- flat rotation curves of all spiral galaxies
- Tully Fisher's Law $M \sim v_0^4$
- it provides a criteria for the scale at which the dynamics deviates from Newton's Law, the surface mass density drops beneath $\Sigma_m = a_0/G$. Because luminosity tracks mass, we expect that low surface luminosity galaxies to present deviations of Newtonian dynamics. Indeed, for LSB galaxies which are galaxies composed mostly of hydrogen and not stars, the surface gravity falls bellow Σ_m all the way from the bulge to its outskirts and the galaxy displays blatant deviations from Newton's law everywhere.

  It is Important to emphasize that the acceleration v_0^4/GM agrees with a_0 for hundreds of galaxies. There is no a priori reason for this agreement, unless there is some new dynamical law. It is still possible to reproduce Tully's Fisher law from dark matter paradigm but it requires juggling skills (again, fine tuning parameters).

Milgrom's paradigm which is simpler, more economical, falsifiable with a single parameter. Nevertheless, it cannot be the ultimate theory, for many reasons, we mention a few:

- It is ambiguous: Consider the gas inside a cluster where the gravitational field is large enough to produce large accelerations but the cluster itself is immersed in a region where accelerations are small (both with respect to a_0). Which limit of MOND should one consider?
- It does not lead to momentum or angular momentum conservation even if space is homogeneous and isotropic. Consider two masses interacting gravitationally and say $m_2 >> m_1$ such that m_1 behaves Newtonially while m_2 remains in the Mond regime. Clearly $\dot{p} = m_1 a_1 + m_2 a_2$ will not vanish .
- it cannot be generalized into the relativistic regimes. We need a relativistic theory to address gravitational lensing effects, structure formation, cosmological expansion,...

With regard to the first two issues, it would be desirable to frame a modification of Newton dynamics in form of a variational principle (conservation laws are implemented automatically) and change the perspective of modifying inertia, to that of modifying the gravitational fields. The first attempt which is a non-relativistic theory was done by Milgrom himself and Bekenstein in the form of a non quadratic Lagrangian (AQUAL) [16] . They postulated the Lagrangian:

$$S_{\rm grav} = -\int \frac{a_0^2}{8\pi G} F\left[(\nabla\Phi)^2/a_0^2\right] d^3x + \int \rho\Phi d^3x \tag{7.3}$$

Field equations are provided by the variations:

$$\nabla \cdot \left(F'\left[(\nabla\Phi)^2/a_0^2\right] \nabla\Phi\right) = 4\pi G\rho \tag{7.4}$$

where $F'(x) = dF/dx$ The particle moves under the influence of the potential Φ that is $\vec{a} = -\nabla\Phi$ and not the Newtonian potential $\nabla^2\Phi_N = -4\pi G\rho$. Clearly:

$$\nabla \cdot \left(\mu\left[\frac{a}{a_0}\right]\vec{a}\right) = \nabla\Phi_N \tag{7.5}$$

where $\mu(x) = F'(x^2)$. In order to be precise one has to add to the rhs of this expression the curl of some vector field. In spherical symmetrical situations this vector field has to be irrotational and can be omitted. How does AQUAL compare with observations?

- Since it reproduces MOND dynamics, rotation curves are well explained.
- It is prone to instabilities. Actually N-body numerical simulations shows that bars are formed. The long term bar strength distribution resemble the observed pattern and the outcome of the collapse is similar to the observed profiles.
- The time scale for the merger of galaxies is reduced with respect to Newtonian theory with dark matter halo. This is consistent with the evidence that they merge fast.
- AQUAL does not solve the acceleration discrepancy in clusters of galaxies (the same was true for MOND)

In summary, AQUAL does better than dark matter in a galactic scale, with the exception of the times of globular clusters. It fails on the cluster of galaxies scale. Being non relativistic, it cannot address cosmological issues or gravitational lensing.

7.5 Towards a relativistic theory

RAQUAL was the first relativistic generalization of AQUAL [16]. The physical metric $g_{\mu\nu}$ is related to the Einstein metric $\tilde{g}_{\mu\nu}$, the one that satisfies Einstein's equations, via a conformal factor

$$g_{\mu\nu} = e^{2\Phi}\tilde{g}_{\mu\nu} \tag{7.6}$$

The scalar field action mimics AQUAL;

$$L_{\Phi} = -\frac{\eta}{2G}\int F\left(\Phi_{,\alpha}\Phi_{,\beta}g^{\alpha\beta}\right)\sqrt{-g}d^4x \tag{7.7}$$

where η is some dimensionless constant and F is the same function that shows up in AQUAL.

In the weak field limit in GR $g_{00} \approx -1 + 2\Phi_N$; in RAQUAL , further expanding the exponential $\tilde{g}_{00} \approx -1 + 2\Phi_N + 2\Phi$. Thus, whenever gravity is strong Φ_N wins over Φ and we recover Newtonian Physics. In the opposite limit, we recover AQUAL. There is no problem with the (weak) equivalence principle, as we can always solve the differential equations that lead to vanishing Christoffel symbols of the physical metric in a given neighborhood. This naive theory is plagued with problems. First, it cannot predict gravitational lensing since a conformal transformation does not change the light cone structure (the indirect effects via the energy momentum tensor of the scalar field are rather small). Furthermore, there is the causality issue, the scalar field propagates superluminally.

Sanders was the first to realize that in order to predict light bending that differs significantly from GR there is the need to introduce a constant time like vector field U_α, that contracts the space-hypersurfaces orthogonal to this field $e^{-2\phi}(g_{\alpha\beta} + U_\alpha U_\beta)$ and expands space time in the direction of this field $e^{2\phi} U_\alpha U_\beta$. In other words, the physical metric is:

$$\tilde{g}_{\alpha\beta} = e^{-2\phi} g_{\alpha\beta} - \left(e^{2\Phi} - e^{-2\Phi}\right) U_\alpha U_\beta \tag{7.8}$$

The theory breaks Lorentz invariance as it takes a preferred direction in space-time. Furthermore , the vector field is non-dynamical something foreign to the spirit of field theory. Bekenstein promoted U_α to a dynamical field via

$$S_{vec} = -\frac{K}{4G} \int F^{\mu\nu} F_{\mu\nu} \sqrt{-g} d^4x, \qquad F_{\mu\nu} \equiv \partial_\mu U_\nu - \partial_\nu U_\mu \tag{7.9}$$

and the time-like property is enforced via a Lagrange multiplier in the Lagrangian

$$S_{constr} = \lambda \int (U_\alpha U^\alpha + 1) \sqrt{-g} d^4x \tag{7.10}$$

This action defines TeVeS (Tensor + Vector + Scalar field). The scalar field evolves according to equation (7.7). In a twist of the theory, the ugly non-quadratic action can be replaced by a quadratic one coupled to some additional non-dynamical scalar field σ plus some potential $F(\sigma)$:

$$S_\Phi = -\frac{1}{2} \int \left[\sigma^2 h^{\mu\nu} \Phi_{,\mu} \Phi_{,\nu} + \frac{G}{2L^2} \sigma^4 F(kG\sigma^2) \right] \sqrt{-g} d^4x \tag{7.11}$$

where $h^{\mu\nu} = g^{\mu\nu} - U^\mu U^\nu$. The theory reduces to AQUAL theory in the weak field limit, thus reproduces all the results at the galactic scale. The challenge of the theory is to go beyond the galactic scale, namely, cosmology and gravitational lensing.

- In the strong lensing, light deflected by a point-like source, will behave as GR provided that the impact parameter $b << b_0 = k\sqrt{GM/a_0}$. For large impact parameters $b >> b_0$, the deflection angle attains a constant value. More interesting from the observational point of view is the deection of light from a quasar by an intervening galaxy. Taking the Hernquist distribution and spherical model it was possible to show that the galaxy masses are consistent with image positions and amplifications in TeVeS!. The angular separation of images for quasars is better explained in TeVeS (with a cosmological constant) than it does with DM for a large sample of quasars.

- In weak lensing (gravity is not strong enough to split the source into images) the lensing can be obtained through MOND arguments, the deflection angle of a light ray with impact parameter b from a gravitational source M
$$\phi = 2b\int_{-\infty}^{\infty} dl\frac{a(r)}{r} \tag{7.12}$$
where l is the length over the photon trajectory and r is the distance between the photon and the CM of the mass distribution. The gravitational acceleration $a(r)$ is given by MOND's formula
$$\mu\left(\frac{a}{a_0}\right) = \frac{M(< r)}{r^2} \tag{7.13}$$
where $M(< r)$ is the mass circumscribed with by the sphere of radius r. Unfortunately, MOND cannot account for the observations, there is a need of a dark halo of neutrinos of mass range $2 - 6\ eV$ for many galaxy clusters. A full TeVeS lensing calculation does not relieve: the bullet cluster lensing cannot accounted unless one adds dark halo in the form of neutrinos, which are already excluded.

- Last, but not least, Cosmology. In the formalism of density perturbations in TeVeS, fluctuations in the vector field sector are responsible for amplification of density perturbations even if the energy-momentum tensor of the additional fields (scalar+vector) remains small $\Omega_X << 1$ [26]. In other words, the amplification of the perturbations in TeVeS does not arise from the contribution of the additional fields to the energy momentum ("dark matter" feature) rather from the structure of the theory itself.

TeVeS accounts properly for the pattern galaxies and the multipole distribution of the background microwave radiation. On this direction, recent work seems to show that the scalar field acting as dark matter can reproduce the observed accelerated expansion and inflation at an early epoch. So it might be that a modified theory could overcome the need of a dark matter/energy filling up space.

7.6 Variant Theories

There are some variant theories modifying GR in order to reproduce observations without resorting to the dark matter:

- Moffat proposed another variation of GR, MOG (Modified Gravity) [20], which requires the addition of a massive vector field and three scalar fields. The action for the vector field is:

$$S_{vector} = -\frac{1}{4\pi}\int \omega\left[\frac{1}{4}F^{\mu\nu}F_{\mu\nu} - \frac{1}{2}m^2U^{\mu}U^{\nu} + V_U\right]\sqrt{-g}d^4x \quad (7.14)$$

  where V_U is a self-interaction potential for the vector field U^{μ}. In this theory all coupling constants (Newton's constant, the vector field mass and ω) became dynamical fields:

$$\begin{aligned} S_{\omega\omega\varphi} &= -\frac{1}{G}\int [\ln G)^{\alpha}(\ln G)_{\alpha} + (\ln m)^{\alpha}(\ln m)_{\alpha} \\ &+ (\ln\omega)^{\alpha}(\ln\omega)_{\alpha} - V_G - V_m - V_{\omega}]\sqrt{-g}d^4x \quad (7.15) \end{aligned}$$

  Here, V_G, V_m and V_{ω} are the self-interaction terms for the scalar fields G, m and ω. Clearly, solving these equations requires two integration constants for each field. These constants in the theory, are determined in a case by case basis, depending on the kind of problem we are dealing with (weak gravitational lensing, rotation curves in a galaxy....) in order to reproduce observations. The myriad of integration constants fine tuned in one instance cannot be exported to another - the fine tuning is made in a case by case basis. That's makes the theory less appealing.

- Zlosnik, Ferreira and Starkman proposed a theory (Aether Theory) based on a single time-like vector field. The theory has a standard Einstein action while the vector field, which does not couple directly to matter, has the action

$$L = \frac{M^2}{16\pi G}\mathcal{F}(\mathcal{K}) + \frac{\lambda}{16\pi G}(A^\mu A_\mu + 1) \tag{7.16}$$

where λ is a Lagrange multiplier that enforces the vector field time-likeness, M is a parameter of the order of a_0 , $\mathcal{F}$ is some (yet unspecified function) and

$$\mathcal{K} = M^{-2}K^{\alpha\beta}_{\mu\nu}\nabla_\alpha A^\mu \nabla_\beta A^\nu \tag{7.17}$$

$$K^{\alpha\beta}_{\mu\nu} = c_1 g^{\alpha\beta} g_{\mu\nu} + c_2 \delta^\alpha_\mu \delta^\beta_\nu + c_3 \delta^\alpha_\nu \delta^\beta_\nu \tag{7.18}$$

where c_1, c_2 and c_3 are constants of the theory. In an extension of the theory [22] an extra term $-c_4 A^\alpha A^\beta g_{\mu\nu}$ is added to $\mathcal{K}^{\alpha\beta}_{\mu\nu}$.

In contrast with TeVeS, the non-quadratic Lagrangian rests now on the vector field. The theory enhances cosmological density perturbations similar to TeVeS [21], which is very welcome.

In the weak field limit where $h_{00} \equiv -2\Phi$ (Φ is the gravitational potential) and $A_0 \equiv -1 + B$, the theory can be solved linearly (including the constraint) and it follows that

$$\nabla \cdot [(2 + c_1\mathcal{F}')\nabla\Phi] = 8\pi G\rho \tag{7.19}$$

MOND dynamics is exactly reproduced if $\mathcal{F} = a\mathcal{K} + b\mathcal{K}^{3/2}$ for $\mathcal{K} << M$. The theory can explain the accelerated expansion of the Universe for a wide class of functions $\mathcal{F}$ without resorting to a cosmological term. At the solar system all Parameterized post-Newtonian coefficients are reproduced for the particular choice $c_4 = -c_3^2/c_1$ and $c_2 = (-c_1^2 - c_1 c_3 + c_3^2)/3c_1$ [23]. Further constraints on the parameters follow from He^4 cosmological abundance [23] and the maximum observed mass of neutron star [24], [25].

- Sorkis' Generalized TeVeS is a bimetric theory like TeVeS with a Einstenian metric g and a physical metric g_p related by the very same relation as TeVeS (equation (7.8)). The vector and scalar fields actions are:

$$S_A = -\frac{1}{16\pi G}\int\left[K^{\alpha\beta\mu\mu}A_{\mu;\alpha}A_{\nu;\beta} - \lambda(A^\mu A_\mu + 1)\right]\sqrt{-g}d^4x \quad (7.20)$$

$$S_\Phi = -\frac{1}{16\pi G}\int\left[(\mu g^{\mu\nu} - A^\mu A^\nu)\,\Phi_{,\mu}\Phi_{,\nu} + V(\mu)\right]\sqrt{-g}d^4x \quad (7.21)$$

Here $V(\mu)$ is some potential for the non-dynamical field μ, λ a Lagrange multiplier and

$$\begin{aligned} K_{\alpha\beta mu\nu} &= c_1\left(g_{\alpha\mu}g_{\beta\nu} - g_{\alpha\nu}g_{\beta\mu}\right) + c_2\left(g_{\alpha\mu}g_{\beta\nu} + g_{\alpha\nu}g_{\beta\mu}\right) \\ &+ c_3 g_{\alpha\beta}g_{\mu\nu} + c_4 g_{\mu\nu}A_\alpha A_\beta \end{aligned} \quad (7.22)$$

The potential is chosen is such a way that $V(\mu) \sim \mu^3$ in order to reproduce MOND dynamics. As for the Aether theory, matter couples to the gravitational field but not to the vector.

Important to remark that in all variations of the TeVeS (original, Aether and Sorkis) the vector field sector develops instabilities, it is prone to the formation of caustics . This can be easily seen in all the versions of the theory where the dynamics is purely Maxwellian ($F_{\mu\nu}F^{\mu\nu}$) and there is no coupling of the vector field to the matter sector: consider fluctuations P^μ such that $F_{\mu\nu} = 0$ and $P^\mu P_\mu = -1$. With this choice, the field equations remain unchanged. The P^μ field can be easily shown to correspond to time-like geodesics. Thus, for a static star, one can take as an initial condition a inward field, the integral lines of the field are the free falling trajectories. As a consequence , these lines will intersect (caustics) in a free falling time scale. At these points the field is ill defined. An argument involving the Raychaudhuri equation where the geodesic convergence rate $\theta = \vec{\nabla}\cdot\vec{A}$ shows the divergence of θ diverges in a finite proper time in pure TeVeS. In order to circumvent this hurdle , a term $(\vec{\nabla}\cdot\vec{A})^2$ is added in more recent versions of the theory.

7.7 Question Marks

At the outskirts of the solar system the spacecrafts Pioneer 10 and 11 felt both an anomalous acceleration $(8.74 \pm 1.33) \times 10^8 cm/s^2$ [28]. The cause may be mundane, but the coincidence is puzzling. Cassini also reported gravitational anomaly at three times that value. It seems that happens at the MOND scale!

On the other hand, in a recent paper Reyes *et al.* [30] compared red-shifts and gravitational (weak) lensing of background galaxies, following the method put forward by Zhang [29] to differ between Λ CDM cosmology and modifications of gravity. They regarded 70,000 galaxies from the Sloan Digital Sky Survey and their work does favor standard cosmology, at least with regard to the original TeVeS theory.

No matter how the conundrum is settled, it clearly would represent a breakthrough in the understanding of the Universe. Should a new theory of gravity emerge as the favorite, then we are challenged to find a guiding physical principle behind the theory, should the dark matter-dark energy paradigm win, them we will be led with the challenge of identifying the origin of the dark matter-dark energy. The dark energy conundrum could well be settled in the LHC particle's accelerator. The lesson we can learn from Vernier is that one should be open minded for all possibilities.

Bibliography

[1] F. Zwicky,"On a New Cluster of Nebulae in Pisces," In: Proc. Natl. Acad. Sci. U S A., May 1937, vol. 23(5), pp. 251256.

[2] L.M.J.S. Volders, "Neutral Hydrogen in M33 and M101," *Bull. astr. Inst. Netherl.*, vol. 14, 323, 1959. V.C. Rubin, W.K. Ford Jr., N. Thonnard, and M.S. Roberts, "Motion of the Galaxy and the Local Group Determined from the Velocity Anisotropy of Distant Sc I Galaxies. I. The Data and II. The Analysis for the Motion," *Astrophys. J.*, vol. 81, 687 and 719, 1976.

[3] V.C. Rubin, and W.K. Ford Jr., "Rotation of the Andromeda Nebula from a Spectroscopic Survey of Emission Regions," *Astrophys. J.*, vol. 159, 379, 1970.

[4] V.C. Rubin, W.K. Ford Jr., and N. Thonnard, "Rotational Properties of 21 Sc Galaxies with a Large Range of Luminosities and Radii from NGC 4605 (R=4kpc) to UGC 2885 (R=122kpc)," *Astrophysical Journal*, vol. 238, 471, 1980.

[5] R.B. Tully, and J.R. Fisher, "A new method of determining distances to galaxies," *Astron. and Astroph.*, vol. 54, no. 3, 661-673, 1977.

[6] S.M. Faber, and R.E. Jackson, "Velocity dispersions and mass-to-light ratios for elliptical galaxies," *Astroph. J.*, vol. 204, 668-683, 1976.

[7] A.N. Taylor,, S. Dye,, T.J. Broadhurst, N. Benitez, and E. van-Kampen, "Gravitational Lens Magnification and the Mass of Abell 1689," *Astroph. J.*, vol. 501, 1998. Available: doi:10.1086/305827, Los Alamos Archive astro-ph/9801158.

[8] D. Clowe, M. Bradac, A.H. Gonzalez, M. Markevitch, S.W. Randall, C. Jones, D. Zaritsky,, "A Direct Empirical Proof of the Existence of Dark Matter," *Astroph. J. Lett.*, 648, 109-113, 2006.

[9] M. Bradac, D. Clowe, A.H. Gonzalez, P. Marshall, W. Forman, C. Jones, M. Markevitch, S.W. Randall, T. Schrabback, D. Zaritsky,, "Strong and Weak Lensing United. III. Measuring the Mass Distribution of the Merging Galaxy Cluster 1ES 0657-5581," *Astrophys. J.*, vol. 652, 937-947, 2006.

[10] M. Bradac, S.W. Allen, T. Treu, H. Ebeling, R. Massey, R.G. Morris, A. von der Linden, and D. Applegate, "Revealing the Properties of Dark Matter in the Merging Cluster MACS J0025.41222," *Astrophy. J.*, vol. 687, 959, 2008.

[11] A.H. Guth, "The Inflationary Universe: A Possible Solution to the Horizon and Flatness Problems", *Phys. Rev. D*, vol. 23, 347, 1981.

[12] G. Hinshaw, J.L. Weiland, R.S. Hill, N. Odegard, D. Larson , C.L. Bennett, J. Dunkley, B. Gold, M.R. Greason, N. Jarosik, E. Komatsu, M.R. Nolta, L. Page, D.N. Spergel, E. Wollack, M. Halpern, A. Kogut, M. Limon, S.S. Meyer, G.S. Tucker, and E.L. Wright, "Five-Year Wilkinson Microwave Anisotropy Probe Observations: Data Processing, Sky Maps, And Basic Results," *The Astrophysical Journal Supplement*, vol. 180, 225-245, 2009.

[13] M.S. Turner, and D. Huterer, "Cosmic Acceleration, Dark Energy, and Fundamental Physics," *J. Phys. Soc. Jpn.*, vol. 76, 111015, 2007.

[14] J.P. Ostriker, and P.J.E. Peebles, "A Numerical Study of the Stability of Flattened Galaxies: or, can Cold Galaxies Survive?," *Astroph. J.*, vol. 186, 467, 1973.

[15] M. Milgrom, "A modification of the Newtonian dynamics as a possible alternative to the hidden mass hypothesis," *Ap. J.*, vol. 270, 365, 1983.

[16] J.D. Bekenstein and M. Milgrom, "Does the missing mass problem signal the breakdown of Newtonian gravity?," *Ap. J.*, vol. 286, 7, 1984.

[17] R.H. Sanders, "A Stratified Framework for Scalar-Tensor Theories of Modified Dynamics," *Ap. J.*, vol. 480, 492, 1997.

[18] J.D. Bekenstein, "Relativistic gravitation theory for the modified Newtonian dynamics paradigm," *Phys. Rev. D*, vol. 70, 083509, 2004.

[19] C. Skordis, D.F. Mota, P.G. Ferreira, and C. Boehm, "Large Scale Structure in Bekensteins Theory of Relativistic Modified Newtonian Dynamics", *Phys Rev. Lett.*, vol. 96, 011301, 2006.

[20] J.W. Moffat, "Modified Gravitational Theory and Galaxy Rotation Curves," Available: arXiv gr-qc/0404076v4 [2004].

[21] T.G. Zlosnik, P.G. Ferreira, and G.D. Starkman, "Modifying gravity with the aether: An alternative to dark matter," *Phys. Rev. D*, vol. 75, 044017, 2007.

[22] D. Garfinkle, C. Eling and T. Jacobson, "Numerical simulations of gravitational collapse in Einstein-aether theory," *Phys. Rev. D*, vol. 76, 024003, 2007.

[23] T. Tamaki, and U. Myamoto, "Generic features of Einstein-Aether black holes," *Phys. Rev. D*, vol. 77, 024026, 2008.

[24] H. Quaintrell, A.J. Norton, T.D.C. Ash, P. Roche, B. Willems, T.R. Bedding, I.K. Baldry, and R.P. Fender, "The mass of the neutron star in Vela X-1 and tidally induced non-radial oscillations in GP Vel," *Astron. Astrophys.*, vol. 401, 313-323, 2003.

[25] D.J. Nice, E.M. Splaver, I.H. Stairs, O. Lohmer, A. Jessner, M. Kramer, and J.M. Cordes, "A 2.1 M Pulsar Measured by Relativistic Orbital Decay," *Astrophys. J.*, vol. 634, 1242, 2005.

[26] C. Skordis, "Generalizing tensor-vector-scalar cosmology," *Phys. Rev. D*, vol. 77, 123502, 2008.

[27] C.R. Contaldi, T. Wiseman, and B. Withers, "TeVeS gets caught on caustics," *Phys. Rev. D*, vol. 78, 044034, 2008.

[28] J.D. Anderson, and P.A. Laing, E.L. Lau, A.S. Liu, M.M. Nieto, and S.G. Turyshev, "Study of the anomalous acceleration of Pioneer 10 and 11," *Phys. Rev. D*, vol. 65 , 082004, 2002.

[29] P. Zhang, M. Liguori, R. Bean, and S. Dodelson, "Probing Gravity at Cosmological Scales by Measurements which Test the Relationship between Gravitational Lensing and Matter Overdensity," *Phys. Rev Lett.*, vol. 99, 141302, 2007.

[30] R. Reyes, R. Mandelbaum, U. Seljak, T. Baldauf, J.E. Gunn, L. Lombriser, and R.E. Smith, "Confirmation of general relativity on large scales from weak lensing and galaxy velocities," *Nature*, vol. 464, 256-258, 2010.

CHAPTER 8

Continuity Equation for an Embedded System

Robert Englman[1]

Abstract

By a special representation of the number density operator and working with a modulus-phase formalism we obtain the dissipative current, needed to ensure particle number (or charge) conservation in the conducting system coupled to an environment. This generalizes and simplifies previous derivations aimed at the Lindblad type of master equations. In addition to a part depending linearly on the Hamiltonian current, we obtain also a pseudo-curl contribution to the dissipative current.

8.1 Introduction

The theory of quantum mechanics is mathematically speaking a theory of a complex field that must satisfy a complex partial differential equation, the Schrödinger equation. This field is unique in the sense that its modulus square represents a probability and hence must satisfy a continuity type conservation law. This continuity equation can indeed be deduced from the

[1]Department of Physics and Applied Mathematics, Soreq NRC, Yavne 81800, Israel
e-mail: englman@vms.huji.ac.il;

Schrödinger equation when the quantum system is isolated. However, difficulties arise when the quantum system is in contact with the environment. It is thus imperative to describe such an equation when the situation is not ideal as it usually is. Moreover, this is considered an advancement in the field theory of quantum mechanics since it widens the scope of the theory such that it includes non-isolated systems.

Dissipative processes in materials have found their ways to standard textbook, such as [1], which have noted that dispersive media lack a thermodynamic interpretation. (A dispersive solid is by necessity dissipative, in that through the Kramers-Kronig relations its constitutive constants, that are characterized by a frequency dependence, must possess an imaginary part, which represents absorption or dissipation.) At the same time, dissipation requires the availability of a repository of the energy, here named environment (E) in which the charge propagating system (S) is "embedded". It turns out that treatments of the electromagnetic current-field in a dissipative solid contain an unsuspected blemish, and this is the subject of the present work.

It was noticed by Gebauer and Car [2] that when the dissipative effects are incorporated in the quantum master equations in a Markovian approximation, they do not satisfy the equation of continuity (or conservation) involving the particle density $n(\vec{x}, t)$ and the Hamiltonian current $\vec{J}_H(\vec{x}, t)$ (only). Here $\vec{x}$ is the position inside the system (S) being in interaction with its environment (E); t is time. They have exemplified the breakdown with a Hamiltonian H_S acting locally upon the system's coordinates and for rate equations for the system's (reduced) density matrix ρ_S of the Lindblad type [3]. They have further succeeded in formulating a "dissipative" current $\vec{J}_C(\vec{x}, t)$, which, when supplementing the Hamiltonian current, satisfies the continuity equation. Subsequently, a newer derivation of this result was given with the purpose of its being more transparent [4]. The aim of the present work is to present a very short generalization of the previous results beyond the Lindblad-form. It is based on expressing the wave function in a modulus-phase form, which leads to the continuity equation in a straightforward manner [5, 6].

8.2 From the Schrödinger to the Conservation Equation

The state of the total system ($T = S \bigoplus E$) is expressed by a time dependent wave function $\Psi^T(\mathbf{r}_S, \mathbf{r}_E, t)$, in which $\mathbf{r}_S$ denote all the system coordinates and $\mathbf{r}_E$ all the environment coordinates, this being a solution of the closed system Schrödinger equation

$$i\frac{\partial \Psi^T(\mathbf{r}_S, \mathbf{r}_E, t)}{\partial t} = [H_S(\mathbf{r}_S) + H_E(\mathbf{r}_E) + V_I(\mathbf{r}_S, \mathbf{r}_E)]\Psi^T(\mathbf{r}_S, \mathbf{r}_E, t) \qquad (8.1)$$

($\hbar = 1$). The total Hamiltonian (H^T) has been split into system (H_S), environment (H_E) and local interaction (V_I) parts: the former two contain the kinetic energy operators

$$-\sum_s \frac{1}{2m_S}\nabla_s^2, \quad -\sum_e \frac{1}{2m_E}\nabla_e^2 \qquad (8.2)$$

(respectively), where s labels a constituent of the system S (e.g., an electron) and e labels a constituent of the environment E; m_S,m_E are the common masses of the constituents. Expressing the wave function in the modulus-phase form

$$\Psi^T(\mathbf{r}_S, \mathbf{r}_E, t) = |\Psi^T(\mathbf{r}_S, \mathbf{r}_E, t)|e^{i\phi^T(\mathbf{r}_S, \mathbf{r}_E, t)} \qquad (8.3)$$

one obtains the Schrödinger equation in the form of two equations, of which the first is

$$\frac{\partial |\Psi^T(\mathbf{r}_S, \mathbf{r}_E, t)|^2}{\partial t} + \sum_{(S,E)} \nabla_{(S,E)} \cdot \left(|\Psi^T(\mathbf{r}_S, \mathbf{r}_E, t)|^2 \mathbf{v}(\mathbf{r}_S, \mathbf{r}_E, t)\right) = 0 \qquad (8.4)$$

where

$$\mathbf{v}(\mathbf{r}_S, \mathbf{r}_E, t) \equiv \frac{1}{m(S,E)}\nabla_{(S,E)}\phi^T(\mathbf{r}_S, \mathbf{r}_E, t) \qquad (8.5)$$

$m(S, E)$ is an effective mass parameter. There is also a second equation (shown in [5, 6] and equivalent to the Hamilton-Jacobi equation) which is needed to complete the connection between the modulus $|\Psi^T|$ and the phase ϕ^T of the wave function, but we shall make no use of this here.

By introducing the density function of the total system in the coordinate representation as

$$\rho^T(\mathbf{r}_S, \mathbf{r}_E; \mathbf{r}'_S, \mathbf{r}'_E, t) \equiv \Psi^T(\mathbf{r}_S, \mathbf{r}_E, t)\Psi^{T*}(\mathbf{r}'_S, \mathbf{r}'_E, t) \qquad (8.6)$$

In this definition all $\mathbf{r}$ symbols are operators and, therefore, so is ρ^T. We then obtain from equation (8.4) an equation for the diagonal part of the density matrix

$$\frac{\partial \rho_T(\mathbf{r}_S, \mathbf{r}_E; \mathbf{r}_S, \mathbf{r}_E, t)}{\partial t} + \nabla_{(S,E)} \cdot \vec{J}_H^T(\mathbf{r}_S, \mathbf{r}_E, t) = 0 \tag{8.7}$$

in which the $\mathbf{r}$ symbols are coordinate values. The density matrix in the last equation is connected to the density function operator through

$$\rho_T(\mathbf{r}_S, \mathbf{r}_E; \mathbf{r}'_S, \mathbf{r}'_E, t) =< \mathbf{r}_S, \mathbf{r}_E | \rho_T(\mathbf{r}_S, \mathbf{r}_E; \mathbf{r}_S, \mathbf{r}'_E, t) | \mathbf{r}'_S, \mathbf{r}'_E > \tag{8.8}$$

In equation (8.7) one notes the Hamiltonian current of the total system $\vec{J}_H^T$ defined as:

$$\vec{J}_H^T(\mathbf{r}_S, \mathbf{r}_E, t) = |\Psi^T(\mathbf{r}_S, \mathbf{r}_E, t)|^2 \mathbf{v}(\mathbf{r}_S, \mathbf{r}_E, t) \tag{8.9}$$

It is elementary to deduce from equation (8.7) a conservation equation for the total (closed) system, but we are after a conservation equation for the system exclusive of the environment and expressed in terms of the reduced (or environment-traced) density matrix $\rho_S(\mathbf{r}_S, \mathbf{r}'_S, t)$ for the system only. When carried out, the environment-tracing turns equation (8.7) into

$$\frac{\partial \rho_S(\mathbf{r}_S, \mathbf{r}_S, t)}{\partial t} + i[H_S, \rho_S(\mathbf{r}_S, \mathbf{r}_S, t)] - C(\mathbf{r}_S, t) = 0 \tag{8.10}$$

where the system commutator [,] (appearing also in eq. (2) of [2] and in eq. (1) of [4]) has been introduced, and a new quantity is defined by

$$C(\mathbf{r}_S, t) = -Tr_E(\nabla_{(S,E)} \cdot \vec{J}_H^T) + i[H_S, \rho_S(\mathbf{r}_S, \mathbf{r}_S, t)] \tag{8.11}$$

(Compare this to the quantities denoted by $C[\bar{S}(t)]$ in [2] and by $\pounds\rho$ in [4])

The (coarse-grained) particle density in the system is given [7] by:

$$n(\vec{x}, t) = (\Pi_s \int_V d\mathbf{r}_s) \rho_S(\mathbf{r}_S, \mathbf{r}_S, t) \sum_{s'} \delta(\vec{r}_{s'} - \vec{x}) \tag{8.12}$$

in which $(\Pi_s \int_V d\mathbf{r}_s)$ signifies a multiple integration over the system volume for all system coordinates $\mathbf{r}_s$. (In this "shorthand" expression Π_s is the product symbol. $\rho_S(\mathbf{r}_S, \mathbf{r}_S, t)$ is the matrix element diagonal in **all** system coordinates. The s' summation also runs over all system coordinates. On the other hand, $\mathbf{r}_S$ in $C(\mathbf{r}_S, t)$ and in $\rho_S(\mathbf{r}_S, \mathbf{r}_S, t)$ represents the system coordinates generically.) The second term in equation (8.10) gives the divergence of the Hamiltonian current $\vec{J}_H^S$ for the system [2, 4]. Treating the last term in that equation requires the multiple integration:

$$D(\vec{x}) \equiv (\Pi_s \int_V d\mathbf{r}_s) C(\mathbf{r}_S, t) \sum_{s'} \delta(\vec{r}_{s'} - \vec{x}) \tag{8.13}$$

This is carried out by noting that

$$\delta(\vec{r}) = -\frac{1}{4\pi} \vec{\nabla} \cdot \vec{\nabla} \frac{1}{|\vec{r}|} \tag{8.14}$$

which leads immediately to

$$D(\vec{x}) = \frac{1}{4\pi} \vec{\nabla}_{\vec{x}} \cdot (\Pi_s \int_V d\mathbf{r}_s) C(\mathbf{r}_S, t) \sum_{s' \epsilon S} \vec{\nabla}_{\vec{r}_{s'}} \frac{1}{|\mathbf{r}_{s'} - \vec{x}|} \tag{8.15}$$

Since this is a divergence term (in which the environment coordinates have been traced, integrated out), we can write our conservation equation as

$$\frac{\partial n(\vec{x})}{\partial t} + \vec{\nabla}_{\vec{x}} \cdot (\vec{J}_H^S(\vec{x}) + \vec{J}_D(\vec{x})) = 0 \tag{8.16}$$

where

$$\vec{J}_D(\vec{x}) \equiv -\frac{1}{4\pi} (\Pi_s \int_V d\mathbf{r}_s) C(\mathbf{r}_S, t) \sum_{s' \epsilon S} \vec{\nabla}_{\vec{r}_{s'}} \frac{1}{|\vec{r}_{s'} - \vec{x}|} \tag{8.17}$$

$\vec{J}_H^S$ comes from the system commutator term in equation (8.10) . After some manipulation the middle term (representing the divergence of the Hamiltonian current) in equation (8.16) can be written as

$$\vec{\nabla}_{\vec{x}} \cdot \vec{J}_H^S \equiv (\Pi_s \int_V d\mathbf{r}_s) [\sum_{s' \epsilon S} \delta(\vec{r}_{s'} - \vec{x}) \nabla_{\vec{r}_{s'}}] \cdot \vec{J}_H^S(\mathbf{r}_S, t) \tag{8.18}$$

Our equation (8.17) is the sought for expression (valid in a general situation) for the additional current (termed in [2, 4] the "dissipative current") containing only the system coordinates. Its specific form depends on the way that the environment variables are decoupled from the system coordinates.

8.3 Connection with the Decoupling

The evaluation of the $C(\mathbf{r}_S, t)$ term in equation (8.10) requires some approximative decoupling procedure, including also the satisfaction of the "Hamilton -Jacobi equation". In the Markovian approximation the Lindblad-equations have been derived under restrictions on the time parameters (namely: the environment correlation time << the observation time of the system or the inverse system energy difference << the system's decay time) [8]. Then the quantity $C(\mathbf{r}_S, t)$ is simply the coordinate-diagonal part of the dissipative correction to the Hamiltonian master equation, namely

$$C(\mathbf{r}_S, t) = \sum_i (\hat{V}_i^\dagger \rho_S(t) \hat{V}_i - \hat{V}_i \hat{V}_i^\dagger \rho_S(t))_{\mathbf{r}_S, \mathbf{r}_S} g_i + Hermitian\; conjugate \quad (8.19)$$

where the $\hat{V}$'s are operators (matrices) and g's are scalar strength symbols. Substituting this quantity into equation (8.17) yields the "dissipative current" required for particle (or charge) conservation. Likewise, any other form of $C(\vec{r}_S, t)$ chosen to control the dynamics in a dissipating system will also yield via the same equation the corresponding "dissipative current".

8.4 Ignoring the curl-current.

The dissipative current in equation (8.17) is defined only up to an arbitrary curl-term. It is possible to show that under some circumstances the current is curl-free, i.e., of only a grad form. This can be seen from the way that the position coordinate $\mathbf{x}$ appears in the dissipative current expression of Gebauer and Car, Eq. (16) in [2], namely through

$$\hat{\mathbf{J}}(\mathbf{x})_{kl} \equiv \int d\mathbf{r} \phi_k^*(\mathbf{r}) [\nabla_r \delta(\mathbf{x} - \mathbf{r}) + \delta(\mathbf{x} - \mathbf{r}) \nabla_r] \phi_l(\mathbf{r}) \quad (8.20)$$

We rewrite the right hand side as

$$\int d\mathbf{r} \phi_k^*(\mathbf{r}) [-\nabla_x \delta(\mathbf{x} - \mathbf{r}) + \delta(\mathbf{x} - \mathbf{r}) \nabla_r] \phi_l(\mathbf{r}) \quad (8.21)$$

in which the first term is evidently a gradient (irrotational) term. If, further, the function $\phi_l(\mathbf{r})$ is written as a Bloch wave (or a Floquet function)

$$\phi_l(\mathbf{r}) = e^{i\mathbf{k}_l \cdot \mathbf{r}} \sum_n f_l(\mathbf{r} - \mathbf{r_n}) \quad (8.22)$$

where $\mathbf{r}_n$ is the location of the n'th atom in the solid, then for *long* wave length Bloch-waves on s-type atomic states (for which $f_l(\mathbf{r}-\mathbf{r_n}) = f_l(|\mathbf{r}-\mathbf{r_n}|)$)

$$\nabla_r \phi_l(\mathbf{r}) \simeq e^{i\mathbf{k}_l\cdot\mathbf{r}} \sum_n (\mathbf{r} - \mathbf{r_n}) g_l(|\mathbf{r} - \mathbf{r_n}|) \tag{8.23}$$

where g_l is some other function of the distance variables. This is an irrotational (curl-free) function of the integration variable $\mathbf{r}$, and after the integration, which includes the δ-function, the result is also curl-free in $\mathbf{x}$.

Clearly, there will be electronic states not of the form here postulated for $\phi_l(\mathbf{r})$, (cf. the other extreme of free electron wave-function). What the above example shows is that a possible discrepancy between our and previously obtained results for the dissipation current is not due to the neglect of the curl-current. We now turn to the reformulation of our result in the representation of [2].

8.5 Development of the Dissipation Current

The dissipative current of Gebauer and Car [Eq. (16) in [2] and footnote [8] in [4]] is expressed in a system-energy representation. We need to transform our expression in equation (8.17) , written out in the multiple system-coordinate representation, to an energy representation and to perform some manipulations on it. We rewrite equation (8.17) in a trace (Tr) form as

$$-4\pi \vec{J}_D(\vec{x}) = Tr_{\mathbf{r}_S}[C_{\mathbf{r}_S,\mathbf{r}_S} \sum_{S'} \vec{\nabla}_{\vec{r}_{S'}} \frac{1}{|\vec{r}_{S'} - \vec{x}|}] \tag{8.24}$$

in which $C_{\mathbf{r}_S,\mathbf{r}_S}$ has a matrix form (in fact, a diagonal one) in the **generic** multi-dimensional system-coordinate $\mathbf{r}_S$ representation and where the S' summation is over individual coordinates. The time(t)- dependence has also been suppressed for economy. We transform to the energy representation, diagonal for the system Hamiltonian H^S, by means of a unitary transformation

$$A_{k,\mathbf{r}_S} = \phi_k(\mathbf{r}_S) \tag{8.25}$$

where the right hand side is the k'th energy eigen-function (representative of $|k\rangle$, presumed to be non-degenerate and having energies e_k).

In the energy representation the above trace can be written, utilizing trace-invariance properties, as:

$$-4\pi \vec{J}_D(\vec{x}) = \sum_{kl} C_{kl}\langle k| \sum_{S'} \vec{\nabla}_{\vec{r}_{S'}} \frac{1}{|\vec{r}_{S'} - \vec{x}|} |l\rangle \tag{8.26}$$

In the next crucial step we insert inside the Dirac bracket a commutator divided by an energy difference, so as to leave the value of the bracket unchanged.

$$-4\pi \vec{J}_D(\vec{x}) = \sum_{kl} C_{kl}\langle k|(e_k - e_l)^{-1} \sum_{s\in S} [H_S(\mathbf{r}_S), \vec{\nabla}_{\vec{r}_s} \frac{1}{|\vec{r}_s - \vec{x}|}]|l\rangle \ , \ e_k \neq e_l \tag{8.27}$$

Inside the commutator only the kinetic energy operator is effective, with value:

$$- \frac{1}{2m_S}[\sum_{\alpha,s\epsilon S} \partial_s^\alpha \partial_{s\alpha} \ , \sum_{s'\epsilon S} \vec{\nabla}_{\vec{r}_{s'}} \frac{1}{|\vec{r}_{s'} - \vec{x}|}] = -\frac{1}{2m_S} \sum_{\alpha,s\epsilon S} \{(\partial_s^\alpha \partial_{s\alpha} \vec{\nabla}_{\vec{r}_s} \frac{1}{|\vec{r}_s - \vec{x}|})$$

$$+ \ 2(\partial_s^\alpha \vec{\nabla}_{\vec{r}_s} \frac{1}{|\vec{r}_s - \vec{x}|})\partial_{s\alpha}\} \tag{8.28}$$

having reduced the double sum to a single sum. (m_S is the (common) mass of all particles in the system S , $\hbar = 1$; $\alpha = x, y, z$.) In the curly brackets we now consider one single vector component, labelled k. The first term is (with adoption of the summation convention for doubly occurring greek indices)

$$(\partial_s^\alpha \partial_{s\alpha} \partial_s^k \frac{1}{|\vec{r}_s - \vec{x}|}) = \partial_s^k (\nabla_{\vec{r}_s}^2 \frac{1}{|\vec{r}_s - \vec{x}|}) - (\vec{\nabla}_{\vec{r}_s}^2 \frac{1}{|\vec{r}_s - \vec{x}|})\partial_s^k \tag{8.29}$$

in which the first term will be recognized (casting our view back to equation (8.14) and equation (8.20)) as the first half in the current operator expression.

We now manipulate the second term on the right hand side of equation (8.28) to obtain the second half of the current together with an extra term (resembling a curl-expression). Carrying this out with a shortened notation in which the s (system-component) labels are omitted,

$$\begin{aligned} &2(\partial^\alpha \partial^k \frac{1}{|\vec{r}_s - \vec{x}|})\partial^\alpha \\ &= 2\{(\nabla^2 \frac{1}{|\vec{r}_s - \vec{x}|})\partial^k + \partial^k[(\partial^\alpha \frac{1}{|\vec{r}_s - \vec{x}|})\partial^\alpha] - \partial^\alpha[(\partial^\alpha \frac{1}{|\vec{r}_s - \vec{x}|})\partial^k]\}. \end{aligned} \tag{8.30}$$

(Due to the factor 2) the first term in the above equation cancels the negative part in equation (8.29) and complements the positive part to give $\partial_s^k(\nabla^2_{\vec{r}_s}\frac{1}{|\vec{r}_s-\vec{x}|})+(\vec{\nabla}^2_{\vec{r}_s}\frac{1}{|\vec{r}_s-\vec{x}|})\partial_s^k$, which is just -4π times the Hamiltonian current-density operator. Then, in a vector notation, the curly brackets become, very simply,

$$-4\pi\hat{J}_H(\vec{x})+2\sum_s\vec{M}_s(\vec{x}) \tag{8.31}$$

This gives for the dissipation current in equation (8.27) the result

$$\vec{J}_D(\vec{x})=\sum_{lk}(e_k-e_l)^{-1}C_{lk}\langle k|[\hat{J}_H(\vec{x})-(2\pi)^{-1}\sum_s\vec{M}_s(\vec{x})]|l\rangle \tag{8.32}$$

in which the effect of the environment enters through the matrix elements C_{lk} expressed in the system's (S) energy representation. (Its form in the system coordinate representation is shown in equation (8.11)) The second term contains the operator $\vec{M}_s(\vec{x})$, defined in equation (8.31) .

8.5.1 Relation to previous work

The expressions of [2] and [4] for the dissipation current were linear functionals of the Hamiltonian current. The first term in equation (8.32) is indeed of this form; but not the second term involving $\vec{M}(\vec{x})$. Moreover, this term is not a "curl" term in the field variable ($\vec{x}$) and its divergence does not in general vanish. Thus, our expression (obtained in an easily checkable manner) differs from those of [2] and [4].

8.6 Discussion

The replacement shown in equation (8.14) has enabled us to derive remarkably simply the "dissipation current" needed for particle conservation. The current is defined only up to an arbitrary curl-term. In two dimensions the inverse radius in equation (8.17) is to be replaced by the log of the radius.

Finally, although the particle conservation has been demonstrated, previously in [2, 4] and in the present communication for the general case , an understanding of the flow involved in the "dissipation current" is desirable.

Acknowledgements

Thanks are due to L. Diosi and A. Bodor for helpful remarks.

Bibliography

[1] L.D. Landau, E.M. Lifshitz and L.P. Pitaevskii, Electrodynamics of Continuous Media, Second Edition. Pergamon: Oxford, 1984. Chapter IX.

[2] R. Gebauer, and R. Car, "Current in Open Quantum Systems," *Phys. Rev. Lett.*, vol. 93, 160404, 2004.

[3] G. Lindblad, "On the generators of quantum dynamical semigroups," *Commun. Math. Phys.*, vol. 48, 119-130, 1976.

[4] A. Bodor, and L. Diósi, "Conserved current in Markovian open-quantum systems," *Phys. Rev. A*, vol. 73, 064101, 2006.

[5] P.R. Holland, The Quantum Theory of Motion. Cambridge University Press: Cambridge, 1993.

[6] R. Englman, and A. Yahalom, "Complex States of Simple Molecular Systems," In: The Role of Degenerate States in Chemistry, editors M. Baer, and G.D. Billing, Adv. Chem. Phys., vol. 124, 197 section VI B, John Wiley and Sons: New York, 2002.

[7] L.D. Landau, and E.M. Lifshitz, Quantum Mechanics. Addison-Wesley: Cambridge, Mass., 1951.

[8] W.H. Louisell, Quantum Statistical Properties of Radiation. John Wiley and Sons: New York, 1973. Section 6.2.

CHAPTER 9

The Emergence of non Abelian Gauge Field Theory from the Born - Oppenheimer Treatment

Abstract

The study of non Abelian gauge theories is thought to be in the realm of quantum field theory, having to do mainly with the study of elementary particle phenomena. The non Abelian gauge theory widely known as the "standard model" describes correctly all elementary particles and forces except gravity.

Our goal here is to show that non Abelian gauge theories appear also in physical settings in which the field need not be quantized (second quantization). Hence they can be considered as classical fields which are the subject of this book. In particular it will be shown that those fields appear in the study of molecular systems is the frame work of the Born - Oppenheimer theory.

9.1 Introduction

The study of non Abelian gauge theories is thought to be in the realm of quantum field theory, having to do mainly with the study of elementary particle phenomena. The non Abelian gauge theory widely known as the "standard model" describes correctly all elementary particles and forces except gravity.

Our goal here is to show that non Abelian gauge theories appear also in physical settings in which the field need not be quantized (second quantization). Hence they can be considered as classical fields which are the subject of this book. In particular it will be shown that those fields appear in the study of molecular systems is the frame work of the Born - Oppenheimer theory.

The study of molecular systems is based on the Born-Oppenheimer theory, which distinguishes between rapidly moving electrons and slowly moving nuclei and therefore leads to the formation of the electronic (adiabatic) eigenstates and the corresponding non adiabatic coupling terms (NACTs). The existence of the adiabatic states seems to be compatible with a wide range of experimental studies, from photochemical processes through photodissociation and unimolecular processes and finally to bimolecular interactions accompanied by exchange and/or charge transfer processes. Having available the adiabatic states, the dynamical treatment is frequently carried out applying the Born-Oppenheimer approximation, which assumes the existence of a single decoupled state, thus ignoring the NACTs that couple the various states. The justification for this approximation is in the fact that NACTs are proportional to the mass ratio, namely, $\sqrt{\frac{m_e}{m_p}}$ (where m_e and m_p are the masses of an electron and a proton, respectively) and therefore are expected to be about two orders of magnitude smaller than other characteristic magnitudes that appear in the nuclear Schrödinger equation. However, as more and more molecular systems are studied and the number of ab-initio treatments increases, it is realized that this approximation holds, at most, for small regions in configuration space, but otherwise cannot be satisfied. The main reason for this misfortune is the existence of molecular degenerate states, which in turn cause the NACTs not only to become infinitely large but also to dress as poles. Being poles, the NACTs become the source for numerous phenomena that are considered as topological effects and lead to several interesting issues, including the well-known Longuet-Higgins/Berry phase and the open-path phase, the less known quantization feature of the NACTs, the existence of molecular fields, and finally but not less interesting, topological spin. Since the potential energy surfaces are expected to behave linearly in the vicinity of these points, the degeneracy points are frequently referred to as (points of) conical intersection (cis), the potential energy surfaces may in fact behave parabolically, in the vicinity of a conical intersection, but we still refer to these points as cis. Conical intersections are of most concern because, in contrast to what was believed until the 1990s, they are not rare.

In fact, ab initio calculations indicate that molecular systems tend to form a ci, between every two adjacent states (and on many occasions more than one ci). This fact impacts the numerical study of any process connected with electronic transitions such as energy transfer, chemical reactions, and charge transfer. In other words, unlike in the case of the Born-Oppenheimer approximation, the cis cannot be ignored and the corresponding NACTs, which, because of their singularity and due to the formation of topological effects have to be approached analytically. This analytic approach, which finally leads to the rigorous elimination of the troublesome NACTs (and of the topological effects), is termed diabatization and is one of the more important issues discussed in this book. It is those the NACTs that behave like non-Abelian gauge fields and will be described below.

9.2 Hilbert Space

9.2.1 The Eigenfunction, and the Electronic non-Adiabatic Coupling Term

We consider a complete basis set of electronic eigenfunctions $|\zeta_k(\vec{s}_e|\vec{s})\rangle; k = 1, 2, ..N$, which depend explicitly on the electronic coordinates $\vec{s}_e$ and perimetrically on the nuclear coordinates $\vec{s}$. The $|\zeta_k(\vec{s}_e|\vec{s})\rangle$-functions are the eigenfunctions of the following electronic Hamiltonian $H_e(\vec{s}_e|\vec{s})$:

$$\left(H_e\left(\vec{s}_e|\vec{s}\right) - u_k(\vec{s})\right)|\zeta_k(\vec{s}_e|\vec{s})\rangle = 0 \;\; ; \;\; k = 1, ..., N \tag{9.1}$$

where $u_k(\vec{s})$ are the electronic eigenvalues (that later are recognized as the adiabatic potential energy surfaces (PES)).

The fact that the $|\zeta_k(\vec{s}_e|\vec{s})\rangle$ functions form a complete set yields the resolution of the unity in the following way:

$$I = \sum_{k=1}^{N} |\zeta_k(\vec{s}_e|\vec{s})\rangle\langle\zeta_k(\vec{s}_e|\vec{s})| \tag{9.2}$$

This equation (9.2) guarantees that an arbitrary function $|\xi(\vec{s}_e|\vec{s})\rangle$ can be expressed in terms of a linear combination of the $|\zeta_k(\vec{s}_e|\vec{s})\rangle$-functions with coefficients (which parametrically depend on the nuclear coordinates):

$$|\xi(\vec{s}_e|\vec{s})\rangle = \sum_{k=1}^{N} |\zeta_k(\vec{s}_e|\vec{s})\rangle\langle\zeta_k(\vec{s}_e|\vec{s}) \mid \xi(\vec{s}_e|\vec{s})\rangle \tag{9.3}$$

This expansion is known as the Born-Oppenheimer expansion [1, 3] and the bra-ket notation, introduced by Dirac [4] and used throughout this chapter, implies integration over the electronic coordinates. Next we form the following magnitude:

$$\vec{\tau}_{jk} = \left\langle \zeta_j \,\middle|\, \vec{\nabla}\zeta_k \right\rangle = -\vec{\tau}_{kj}; \qquad k, j = \{1, 2, ..., N\} \tag{9.4}$$

which is defined as the electronic non-adiabatic coupling term (NACT) [1, 3]. Here the grad operator is expressed in terms of the nuclear coordinates $\vec{s}$:

$$\vec{\nabla} = \left\{ \frac{\partial}{\partial p}, \frac{\partial}{\partial q}, ... \right\} \tag{9.5}$$

where p and q are two Cartesian (mass scaled) coordinates. It is important to notice that the elements $\vec{\tau}_{jk}$ form a matrix with elements that are vectors, an alternative view is to think of $\vec{\tau}_{jk}$ as a vector of (anti symmetric) matrices.

With these definitions we form the Hilbert space defined at a given point $\vec{s}$ namely, the space that contains all the functions that solve equation (9.1) at a given point $\vec{s}$ [5]. One of the more important tasks is to connect Hilbert spaces defined at different points, for instance, at $\vec{s}$ and $\vec{s} + \Delta\vec{s}$: It can be shown that a connection is established through the NACTs [6]:

$$|\zeta_k(\vec{s}_e|\vec{s} + \Delta\vec{s})\rangle = \sum_{j=1}^{N} (\delta_{kj} - \Delta\vec{s} \cdot \vec{\tau}_{kj})\, |\zeta_j(\vec{s}_e|\vec{s})\rangle \tag{9.6}$$

where δ_{jk} is the Kronecker delta function and the dot stands for a scalar product. Equation (9.6) is always fulfilled if the Hilbert spaces at point $\vec{s}$ and at points close to it are of N dimensions (namely, contains N eigenfunctions). However, we show later that this relation holds, under certain conditions, also for a smaller group of states that is defined as a Hilbert subspace. To prove equation (9.6) we consider the Taylor expansion for $|\zeta_k(\vec{s}_e|\vec{s} + \Delta\vec{s})\rangle$:

$$|\zeta_k(\vec{s}_e|\vec{s}+\Delta\vec{s})\rangle = |\zeta_k(\vec{s}_e|\vec{s})\rangle + \Delta\vec{s} \cdot \vec{\nabla}\, |\zeta_k(\vec{s}_e|\vec{s})\rangle \tag{9.7}$$

Since the derivative $\vec{\nabla}\,|\zeta_k(\vec{s}_e|\vec{s})\rangle$ is also a function that belongs to the same Hilbert space it can be presented in terms of the (electronic) basis set introduced earlier, thus:

$$\left|\vec{\nabla}\zeta_k(\vec{s}_e|\vec{s})\right\rangle = \sum_{j=1}^{N} \vec{Z}_{kj}(\vec{s})|\zeta_j(\vec{s}_e|\vec{s})\rangle \tag{9.8}$$

Multiplying both sides by $\langle\zeta_i(\vec{s}_e|\vec{s})|$ (which also implies integration over $\vec{s}_e$ yields (see equation (9.4)):

$$\vec{Z}_{ki} = -\vec{\tau}_{ki}; \qquad i,k = \{1,2,....,N\} \tag{9.9}$$

Substituting equations (9.8) and (9.9) in equation (9.7) yields equation (9.6). In what follows we consider only *real* eigenfunctions and for them it is easy to show that the diagonal elements of the $\vec{\tau}$-matrix, namely, $\vec{\tau}_{jj}$ are identically zero since the matrix is anti symmetric. As a result equation (9.6) leads to an important relation known as the *parallel transport law* [7, a] namely:

$$\langle\zeta_k(s_e|s)\mid\zeta_k(s_e|s{+}\Delta s)\rangle = 1 + O(\Delta s^2) \tag{9.10}$$

Equation (9.6) can also be written employing matrix notation:

$$\hat{\zeta}(\vec{s}_e|\vec{s} + \Delta\vec{s}) = \left(\hat{I} - \Delta\vec{s}\cdot\vec{\tau}\right)\hat{\zeta}(\vec{s}_e|\vec{s}) \tag{9.11}$$

where $\hat{I}$ is the unity matrix.

9.2.2 Abelian and the non-Abelian *curl* - Equation

In what follows we distinguish between the Abelian and the non-Abelian magnitudes. The Abelian magnitudes are usually vectors and therefore field operations act on them in the ordinary way as, for instance, in electrodynamics. As an example we may consider the definition of the *curl* equation, for a vector $\vec{\tau}$ which is to be of the form:

$$\vec{H} = Curl\vec{\tau} \tag{9.12}$$

where $\vec{H}$ is also a vector. The *curl* operator implies that the z-component takes the form:

$$H_z = \frac{\partial\tau_y}{\partial x} - \frac{\partial\tau_x}{\partial y} \tag{9.13}$$

The field definition changes in case $\vec{\tau}$ (and consequently also $\vec{H}$) becomes a matrix (of vectors). This situation is known as the non-Abelian case [7, 8] and as will be shown next, the fact that vectors are replaced by matrices of vectors affects relations known to exist between ordinary vector-functions.

As an example we consider the *non-Abelian curl* equation which is sometimes also called as the extended *Curl* equation (to avoid confusion with the ordinary Abelian *Curl* equation). This equation, as mentioned earlier, is written in a matrix form and for this purpose we introduce the following definitions:

(1) Considering two (nuclear) Cartesian coordinates p and q, we define the following tensorial vector $\vec{F}_{pq}$:

$$F_{pq} = \frac{\partial \tau_p}{\partial q} - \frac{\partial \tau_q}{\partial p} - [\tau_p, \tau_q] \tag{9.14}$$

where τ_λ; $\lambda = q, p$ are the λ-components of $\vec{\tau}$, defined as (see equation (9.4)):

$$\tau_{\lambda jk} = \left\langle \zeta_j \left| \frac{\partial}{\partial \lambda} \zeta_k \right. \right\rangle; \quad \{k, j = 1, .., N; \lambda = p, q\} \tag{9.15}$$

and $[\tau_p, \tau_q]$ is the commutation relation between τ_p and τ_q. Equation (9.14) can be written in a more compact way:

$$F = H - T \tag{9.16}$$

where:

$$H_{pq} = \frac{\partial \tau_p}{\partial q} - \frac{\partial \tau_q}{\partial p} \quad \Rightarrow \quad \vec{H} = Curl \vec{\tau} \tag{9.17}$$

and

$$T_{pq} = [\tau_p, \tau_q] \quad \Rightarrow \quad \vec{T} = [\vec{\tau} \times \vec{\tau}] \tag{9.18}$$

F is the field tensor known also as the Yang-Mills field [7],[8]-[12]. Next we prove the following Lemma:

Lemma: For a Hilbert space, F has to be identically zero, namely [13, 14]:

$$F = 0 \tag{9.19}$$

Proof: We consider the p-th and the q-th components of equations (9.17):

$$\left(\frac{\partial \tau_p}{\partial q} \right)_{jk} = \left\langle \frac{\partial \zeta_j}{\partial q} \left| \frac{\partial \zeta_k}{\partial p} \right. \right\rangle + \left\langle \zeta_j \left| \frac{\partial^2 \zeta_k}{\partial q \partial p} \right. \right\rangle \tag{9.20}$$

and

$$\left(\frac{\partial \tau_q}{\partial p}\right)_{jk} = \left\langle \frac{\partial \zeta_j}{\partial p} \,\middle|\, \frac{\partial \zeta_k}{\partial q} \right\rangle + \left\langle \zeta_j \,\middle|\, \frac{\partial^2 \zeta_k}{\partial p \partial q} \right\rangle \tag{9.21}$$

Subtracting equation (9.21) from equation (9.20) and assuming that the eigenfunctions are *analytic* functions with respect to the nuclear coordinates yields the following result:

$$(H_{pq})_{jk} = \left(\frac{\partial}{\partial q}\tau_p - \frac{\partial}{\partial p}\tau_q\right)_{jk} = \left\langle \frac{\partial \zeta_j}{\partial q} \,\middle|\, \frac{\partial \zeta_k}{\partial p} \right\rangle - \left\langle \frac{\partial \zeta_j}{\partial p} \,\middle|\, \frac{\partial \zeta_k}{\partial q} \right\rangle \tag{9.22}$$

Next we consider the (j,k) element of the first term in equation (9.18), namely:

$$(\tau_p \tau_q)_{jk} = \sum_{i=1}^{N} \left\langle \zeta_j \,\middle|\, \frac{\partial \zeta_i}{\partial p} \right\rangle \left\langle \zeta_i \,\middle|\, \frac{\partial \zeta_k}{\partial q} \right\rangle \tag{9.23}$$

and recall the following relation (that holds for real functions):

$$\left\langle \zeta_j \,\middle|\, \frac{\partial \zeta_i}{\partial p} \right\rangle = -\left\langle \frac{\partial \zeta_j}{\partial p} \,\middle|\, \zeta_i \right\rangle$$

Equation (9.23) becomes:

$$(\tau_p \tau_q)_{jk} = -\sum_{i=1}^{N} \left\langle \frac{\partial \zeta_j}{\partial p} |\zeta_i\rangle\langle\zeta_i| \frac{\partial \zeta_k}{\partial q} \right\rangle = -\left\langle \frac{\partial \zeta_j}{\partial p} \middle| \sum_{i=1}^{N} (|\zeta_i\rangle\langle\zeta_i|) \middle| \frac{\partial \zeta_k}{\partial q} \right\rangle$$

or due to the resolution of the unity (see equation (9.2)) we finally obtain:

$$(\tau_p \tau_q)_{jk} = -\left\langle \frac{\partial \zeta_j}{\partial p} \,\middle|\, \frac{\partial \zeta_k}{\partial q} \right\rangle \tag{9.24}$$

A similar result is obtained for the second term in equation (9.18) namely,

$$(\tau_q \tau_p)_{jk} = -\left\langle \frac{\partial \zeta_j}{\partial q} \,\middle|\, \frac{\partial \zeta_k}{\partial p} \right\rangle \tag{9.25}$$

Subtracting equation (9.25) from equation (9.24) yields the following (jk) element of T_{pq}:

$$(T_{pq})_{jk} = (\tau_p \tau_q - \tau_q \tau_p)_{jk} = -\left(\left\langle \frac{\partial \zeta_j}{\partial p} \,\middle|\, \frac{\partial \zeta_k}{\partial q} \right\rangle - \left\langle \frac{\partial \zeta_j}{\partial q} \,\middle|\, \frac{\partial \zeta_k}{\partial p} \right\rangle\right) \tag{9.26}$$

Next, comparing equation (9.26) and equation (9.22), it is well noticed that for any arbitrary components p and q we have:

$$H_{pq} = T_{pq} \quad \Rightarrow \quad F_{pq} = H_{pq} - T_{pq} \; = 0 \tag{9.27}$$

which implies that equation (9.19) is fulfilled for a complete Hilbert space.

As a final issue in this section we consider the two-state ($N = 2$) case. In this case the matrix T is identically zero so that equation (9.16) (see also equation (9.14)) becomes:

$$F = H = Curl\vec{\tau} = 0 \tag{9.28}$$

which due to the following structure of the 2×2 $\vec{\tau}$-matrix:

$$\vec{\tau} = \begin{pmatrix} 0 & \vec{\tau}_{12} \\ -\vec{\tau}_{12} & 0 \end{pmatrix} \tag{9.29}$$

leads to the result that:

$$Curl\vec{\tau}_{12} = 0 \tag{9.30}$$

This result implies that for any pair of coordinates (p, q) we have (in case the Hilbert space is formed by two states):

$$\frac{\partial \tau_{p12}}{\partial q} - \frac{\partial \tau_{q12}}{\partial p} = 0 \tag{9.31}$$

which is the Abelian form of the *curl* equation.

Corollary equation (9.31) indicates that in case of two states ($N = 2$) the Non-Abelian case becomes Abelian. We complete this section by two comments:

(1) We emphasize the fact that all the derivations in this section apply as long as the electronic eigenfunctions are analytic functions at *every* point in the region of interest. In case certain eigenfunctions may not have first order derivatives at some points in configuration space, for instance, at *conical intersections* (see Chapter V of [15]) the respective derivative of the $\vec{\tau}$-matrix elements are not defined at these points and therefore also the final outcome, namely, that F=0 is unlikely to hold at these points.

(2) The importance of the fact that $F = 0$ for Born-Oppenheimer systems and how it is related to other fields in physics is briefly discussed by Englman and Yahalom [10, 11]. In [10] (see especially eq. (70)) the molecular Yang-Mills field F was first introduced in the context of the non-adiabatic coupling term $\vec{\tau}$.

9.2.3 Abelian and non-Abelian *Div*ergence Equations

To guarantee the connection between Hilbert spaces defined at relative remote points i.e. where Δs is relatively large one has to include in equation (9.6) terms, which contain both higher power of $\vec{\tau}_{kj}$ and higher order-derivatives of $\vec{\tau}_{kj}$. In this respect we prove next the relation between the second-order NACTs $\tau_{jk}^{(2)}$ defined as:

$$\tau_{jk}^{(2)} = \langle \zeta_j \mid \nabla^2 \zeta_k \rangle; \quad k, j = \{1, 2, ..., N\} \tag{9.32}$$

and the ordinary (first order) one $\vec{\tau}_{kj}$ introduced in equation (9.4).
Lemma: The second-order NACTs, $\tau_{jk}^{(2)}$ can be presented in the form [13]-[16]:

$$\tau_{jk}^{(2)} = \sum_{i=1}^{N} \vec{\tau}_{ji} \cdot \vec{\tau}_{ik} + \vec{\nabla} \cdot \vec{\tau}_{jk} \tag{9.33}$$

or in matrix notation:

$$\tau^{(2)} = \vec{\tau}^2 + \vec{\nabla} \cdot \vec{\tau} \tag{9.34}$$

Proof:

$$\nabla \cdot \vec{\tau}_{jk} = \nabla \cdot \langle \zeta_j \mid \nabla \zeta_k \rangle = \langle \nabla \zeta_j \mid \cdot \nabla \zeta_k \rangle + \langle \zeta_j \mid \nabla^2 \zeta_k \rangle \tag{9.35}$$

To continue we need to evaluate the first term on the r.h.s of this expression. For this purpose we employ the resolution of the unity given in equation (9.2):

$$\langle \nabla \zeta_j \mid \nabla \zeta_k \rangle = \langle \nabla \zeta_j | \left(\sum_{i=1}^{N} |\zeta_i\rangle\langle\zeta_i| \right) |\nabla \zeta_k\rangle = \sum_{i=1}^{N} \langle \nabla \zeta_j \mid \zeta_i \rangle \langle \zeta_i \mid \nabla \zeta_k \rangle$$

$$= \sum_{i=1}^{N} \vec{\tau}_{ij} \cdot \vec{\tau}_{ik} \tag{9.36}$$

or, finally:

$$\langle \nabla \zeta_j \mid \nabla \zeta_k \rangle = -\sum_{i=1}^{N} \vec{\tau}_{ji} \cdot \vec{\tau}_{ik} \tag{9.37}$$

Substituting equation (9.37) in equation (9.35) (and recalling the definitions given above) yields equation (9.33) (or equation (9.34)).
Corollary

Equation (9.35) was originally derived for the purpose to present the adiabatic nuclear Schrödinger equationin a *complete* and efficient form that guarantees to yield significant physical insight. However it also serves another purpose and for this sake it will be written in a slightly different form, i.e.:

$$\vec{\nabla} \cdot \vec{\tau} = \tau^{(2)} - \vec{\tau}^2 \tag{9.38}$$

equation (9.38) is recognized as the extended *Divergence* equation which together with the extended *Curl* equation (see equations (9.16) - (9.19)) form the *Curl-Div* equations the $\vec{\tau}$-matrix has to fulfill. These two equations are reminiscent of the *Curl-Div* equations for the electromagnetic field tensor in electrodynamics [17, 18], one is reminded that according to a theorem by Helmholtz a vector field is completely defined by its *curl* and *Div.* The two-state system is of special interest because in this case the non-Abelian situation becomes Abelian (namely, the equations for the matrices become equations for the single $(1, 2)$ term), thus in case of the Divergence-equation we have:

$$\vec{\nabla} \cdot \vec{\tau}_{12} = \tau_{12}^{(2)} \tag{9.39}$$

(The two-state $\vec{\tau}^2$-matrix produces only diagonal elements). Equation (9.39) indicates that the Divergence-equation encountered in molecular physics is, in general, different from zero. Equation (9.30) and equation (9.39) form the *curl-Div* equations in case of a two-dimensional Hilbert space.

9.3 Hilbert Subspace

Complete Hilbert spaces, in particular for realistic molecular systems, contain infinite number of states and therefore to be able to treat such systems numerically we have to reveal the conditions or establish situations for which finite groups of states will behave, as a quasi-complete Hilbert space. If such a group of states can be formed we define it as a Hilbert subspace. In the present study the breakup of the Hilbert space into Hilbert subspaces is dictated by the behavior of the previously introduced NACTs elements $\vec{\tau}_{jk}$. It will be shown that the features that exist within in a complete Hilbert space, for instance, the *Cur-Div* equations are *approximately* valid within the respective Hilbert subspace.

In what follows we consider a group of N states and for the sake of convenience we assume them to be the N lowest states. Next we assume the

Hilbert space to break up into two groups – a finite group, designated as the P-space, which contains N (lowest) states, and the complementary group, designated as the Q-space which contains the rest of the Hilbert space (its dimensions can be infinite). As already mentioned the breakup of the Hilbert space is based on the following assumption, namely [19]:

$$|\vec{\tau}_{jk}| \cong O(\varepsilon) \quad for \quad j \leq N; \quad k > N \tag{9.40}$$

Here ϵ is a relatively small number. In other words the NACTs between P-states and Q-states are all assumed to be negligible small. This implies that the $\vec{\tau}$-matrix has the following form [20]:

$$\vec{\tau}=\begin{pmatrix} 0 & \vec{\tau}_{12} & \vec{\tau}_{13} & - & \vec{\tau}_{1N} & & & & & & \\ -\vec{\tau}_{12} & 0 & \vec{\tau}_{23} & - & \vec{\tau}_{2N} & & & & & & \\ -\vec{\tau}_{13} & -\vec{\tau}_{23} & 0 & - & \vec{\tau}_{3N} & & & O(\varepsilon) & & & \\ - & - & - & 0 & - & & & & & & \\ -\vec{\tau}_{1N} & -\vec{\tau}_{2N} & -\vec{\tau}_{3N} & - & 0 & & & & & & \\ & & & & & 0 & \vec{\tau}_{N+1N+2} & \vec{\tau}_{N+1N+3} & - & - & - \\ & & & & & -\vec{\tau}_{N+1N+2} & 0 & \vec{\tau}_{N+2N+3} & - & - & - \\ & & O(\varepsilon) & & & -\vec{\tau}_{N+1N+3} & -\vec{\tau}_{N+2N+3} & 0 & - & - & - \\ & & & & & - & - & - & 0 & - & - \\ & & & & & - & - & - & - & 0 & - \\ & & & & & - & - & - & - & - & 0 \end{pmatrix} \tag{9.41}$$

To continue we define the following Feshbach projection operators, [21, 22] namely, P_N, the projection operator for the P-space:

$$P_N = \sum_{j=1}^{N} |\zeta_j\rangle\langle\zeta_j| \tag{9.42}$$

and Q_N, the projection operator for the Q -space

$$Q_N = I - P_N \tag{9.43}$$

Our aim is to use the Feshbach operators in order to show that both, the extended *curl* equation given in equations (9.16) - (9.19) and the *divergence* equation given in equation (9.38) are, approximately fulfilled within this Hilbert subspace. Since the mathematical proof in both cases is similar we present the proof just for the *Divergence* equation. The proof for the extended *curl* equation can be found in [14]-[24].

Our starting point is, like before, equation (9.33) which, following a slight rearrangement based on equation (9.37), takes the form:

$$\tau_{jk}^{(2)} = -\left\langle \vec{\nabla}\zeta_j \middle| \cdot \vec{\nabla}\zeta_k \right\rangle + \vec{\nabla} \cdot \vec{\tau}_{jk} \tag{9.44}$$

It is important to realize that Eq. (I.39) is valid for any two states j and k and therefore also for those that fulfill j,k $\leq$ N. In what follows we consider only states belonging to the subspace. The first term on the r.h.s is treated further as follows:

$$\langle \nabla\zeta_j \mid \nabla\zeta_k \rangle = \langle \nabla\zeta_j | P_N + Q_N | \nabla\zeta_k \rangle = \langle \nabla\zeta_j | P_N | \nabla\zeta_k \rangle + \langle \nabla\zeta_j | Q_N | \nabla\zeta_k \rangle \tag{9.45}$$

where the first of the two terms in equation (9.45) becomes:

$$\langle \nabla\zeta_j | P_N | \nabla\zeta_k \rangle = \sum_{i=1}^{N} \langle \nabla\zeta_j \mid \zeta_i \rangle \langle \zeta_i \mid \nabla\zeta_k \rangle = \sum_{i=1}^{N} (-\vec{\tau}_{ji}) \vec{\tau}_{ik} \tag{9.46}$$

and the second:

$$\langle \nabla\zeta_j | Q_N | \nabla\zeta_k \rangle = \sum_{i>N} \langle \nabla\zeta_j \mid \zeta_i \rangle \langle \zeta_i \mid \nabla\zeta_k \rangle = \sum_{i>N} (-\vec{\tau}_{ji}) \vec{\tau}_{ik} \tag{9.47}$$

Recalling equation (9.41) it is noticed that the contribution due to equation (9.47) is negligible small (because both $j, k \leq N$) and therefore equation (9.45) becomes:

$$\langle \nabla\zeta_j \mid \nabla\zeta_K \rangle = \langle \nabla\zeta_j | P_N | \nabla\zeta_k \rangle = -\sum_{i=1}^{N} \left(\vec{\tau}_N\right)_{ji} \left(\vec{\tau}_N\right)_{jk} \tag{9.48}$$

which yields for equation (9.44) the following result:

$$\left(\tau_N\right)_{jk}^{(2)} = \sum_{i=1}^{N} \left(\vec{\tau}_N\right)_{ji} \cdot \left(\vec{\tau}_N\right)_{ik} + \vec{\nabla} \cdot \left(\vec{\tau}_N\right)_{jk} \tag{9.49}$$

Equation (9.49) is essentially identical to equation (9.33) which was derived for the complete Hilbert space. Equation (9.49) can be written now also as an equation between the sub-matrices $\vec{\tau}_N$, $\vec{\nabla} \cdot \vec{\tau}_N$ and $\tau_N^{(2)}$:

$$\tau_N^{(2)} = \vec{\tau}_N^{\,2} + \nabla \vec{\tau}_N \tag{9.50}$$

which is similar to equation (9.34). Writing equation (9.50) slightly different namely:

$$\nabla \vec{\tau}_N = \vec{\tau}_N^{(2)} - \vec{\tau}_N^2 \tag{9.51}$$

yields the corresponding *Divergence* equation.

We reiterate that one can show, employing similar considerations, that the curl equations in equation (9.16) - equation (9.19) can be written in the same way for a Hilbert subspace but for the corresponding $N \times N$ sub-matrices.

As the last subject on this issue we would like to estimate the size of the error introduced by ignoring the contributions from the Q-subspace. For this purpose we consider the *curl* equation or, more specifically, the two components of the (unperturbed) vectorial $\vec{\tau}$-matrix, namely, the matrices $\vec{\tau}_q$ and $\vec{\tau}_p$ with elements as presented in equation (9.15). Each of these matrices is now written (following the presentation in equation (9.41)) as follows [24]:

$$\vec{\tau}_x = \begin{pmatrix} \vec{\tau}_x^{(N)} & \vec{\tau}_x^{(N,L)} \\ \vec{\tau}_x^{(L,N)} & \vec{\tau}_x^{(L)} \end{pmatrix} \quad : \quad x = q, p \tag{9.52}$$

Here $\vec{\tau}_x^{(N)}$ and $\vec{\tau}_x^{(L)}$; $x = p, q$ are the diagonal sub-matrices that contain the dominant NACTs whereas $\vec{\tau}_x^{(N,L)}$ (and $\vec{\tau}_x^{(L,N)}$); $x = p, q$ are the two off-diagonal sub-matrices that couple the two diagonal ones and are assumed to contain weak coupling terms, all being of the order $O(\epsilon)$. Next, following equation (9.14), we introduce the components of the reduced tensorial vector $\tilde{F}_{pq}^{(N)}$, namely:

$$\tilde{F}_{pq}^{(N)} = \frac{\partial \vec{\tau}_p^{(N)}}{\partial q} - \frac{\partial \vec{\tau}_q^{(N)}}{\partial p} - \left[\vec{\tau}_p^{(N)}, \vec{\tau}_q^{(N)}\right] \tag{9.53}$$

Recalling equation (9.16)-equation (9.18) the $N \times N$ upper diagonal block of F i.e. $F_{pq}^{(N)}$, can be written in the form:

$$F_{pq}^{(N)} = \tilde{F}_{pq}^{(N)} - \left\{\vec{\tau}_p^{(N,L)} \vec{\tau}_q^{(L,N)} - \vec{\tau}_q^{(N,L)} \vec{\tau}_p^{(L,N)}\right\} \tag{9.54}$$

but due to equation (9.19) we have $F_{pq}^{(N)} = 0$ and therefore we get for $\tilde{F}_{pq}^{(N)}$ the result:

$$\tilde{F}_{pq}^{(N)} = \left\{\vec{\tau}_p^{(N,L)} \vec{\tau}_q^{(L,N)} - \vec{\tau}_q^{(N,L)} \vec{\tau}_p^{(L,N)}\right\} \tag{9.55}$$

Since all four matrices $\vec{\tau}_x^{(N,L)}$ and $\vec{\tau}_x^{(L,N)}$; $x = p, q$ contain elements of the order $O(\epsilon)$ (or smaller) the elements of the matrix on the r.h.s. of equation

(9.54) are of the order $O(\epsilon^2)$ which implies that the relevant elements of $F_{pq}^{(N)}$ are of the order:

$$\tilde{F}_{pq}^{(N)} = O(\varepsilon^2) \tag{9.56}$$

In other words the extended *curl* equation within the Hilbert *subspace*, is fulfilled up to $O(\epsilon^2)$. A similar equation holds for the *Divergence* equation:

$$\vec{\nabla} \cdot \vec{\tau}_N - \tau_N^{(2)} + \vec{\tau}_N^2 = O(\varepsilon^2) \tag{9.57}$$

where $\vec{\tau}_N$ is identical to the previously defined $\vec{\tau}^{(N)}$.

Due to these findings we do not distinguish any more between complete, usually, infinite Hilbert spaces and finite Hilbert subspaces. However we distinguish between groups of states that form a Hilbert subspace and those which do not form a Hilbert subspace.

9.4 The non-Abelian vector potential and its effect on the molecular Schrödinger equation

The Hamiltonian, H, that governs the motion of the nuclei and the electrons is usually written in the following form:

$$H(s_e, s) = T_n(s) + H_e(s_e|s) \tag{9.58}$$

where $T_n(s)$ is the nuclear kinetic energy and $H_e(s_e|s)$ is the electronic Hamiltonian which also contains the nuclear Coulomb interactions and depends parametrically on the nuclei coordinates (the vector arrow $\vec{s}$ was here omitted for brevity). The time independent Schrödinger equationto be considered is of the form:

$$(H - E)\,|\Psi(s_e, s)\rangle = 0 \tag{9.59}$$

where E is the total energy and $|\Psi(s_e, s)\rangle$ is the complete wave function which describes the motion of both, the electrons and the nuclei. Next we employ the Born-Oppenheimer expansion [1, 2, 3]:

$$|\Psi(s_e, s)\rangle = \sum_{j=1}^{N} |\zeta_j(s_e|s)\rangle \psi_j(s) \tag{9.60}$$

where the $\psi_j(s)$, $j = 1, \ldots, N$ are the nuclear-coordinate dependent coefficients (recognized later as the nuclear wave functions) and $|\zeta_j(s_e|s)\rangle; j = 1, \ldots, N$ are the electronic (adiabatic) eigenfunctions of the above introduced electronic Hamiltonian:

$$(H_e(s_e|s) - u_j(s))\,|\zeta_j(s_e|s)\rangle = 0 \;;\quad j = \{1, N\} \tag{9.61}$$

Here $u_j(s), j = \{1, N)$ are the electronic eigenvalues. In this treatment we assume that the Hilbert space is of dimension N. Substituting equation (9.60) in equation (9.59), multiplying it from the left by $\langle\zeta_k(s_e|s)|$ and integrating over the electronic coordinates while recalling equation (9.58) and equation (9.61), yields the following set of coupled equations:

$$\sum_{j=1}^{N} \langle\zeta_k|T_n\psi_j(s)|\zeta_j\rangle + (u_k(s) - E)\,\psi_k(s) = 0 \;;\quad k = \{1, N\} \tag{9.62}$$

Next, we recall that the kinetic operator T_n can be written (in terms of mass-scaled coordinates) as:

$$T_n = -\frac{\hbar^2}{2m}\nabla^2 \tag{9.63}$$

where m is the mass of the system and ∇ is the gradient (vector) operator expressed in terms of mass-scaled coordinates. Substituting equation (9.63) in equation (9.62), performing the corresponding differentiations with respect the nuclear coordinates and the integrations with respect to the electronic coordinates yields the more explicit form of the Born-Oppenheimer system of coupled equations [30]:

$$-\frac{\hbar^2}{2m}\nabla^2\,\psi_k + (u_k - E)\,\psi_k - \frac{\hbar^2}{2m}\sum_{j=1}^{N}\left(2\tau_{kj}\cdot\nabla + \tau_{kj}^{(2)}\right)\psi_j = 0; \qquad k = \{1, N\} \tag{9.64}$$

where τ_{kj} is the (first order) non-adiabatic coupling (vector) matrix with the elements defined in equation (9.4) and $\tau_{kj}^{(2)}$ is the non-adiabatic (scalar) matrix of the second order defined in equation (9.32). Equation (9.64) can also be written in a matrix form as follows [13]:

$$-\frac{\hbar^2}{2m}\nabla^2\Psi + (u - E)\,\Psi - \frac{\hbar^2}{2m}\left(2\tau\cdot\nabla + \tau^{(2)}\right)\Psi = 0 \tag{9.65}$$

where the dot designates the scalar product, $\Psi(s)$ is a column vector that contains the, above mentioned, nuclear functions $\{\psi_j(s), j = 1, \ldots, N\}$ and u is a diagonal matrix which contains the adiabatic potentials. This equation is valid for any group of states.

In Section 9.2 it is proved that for the situation that the group of states forms a Hilbert space (or even a Hilbert sub-space – see Section 9.3) so that the $\vec{\tau}$- matrix fulfills the *divergence* condition (9.34). Substituting equation (9.34) in equation (9.65) yields:

$$-\frac{\hbar^2}{2m}\nabla^2\Psi + \left(u - \frac{\hbar^2}{2m}\tau^2 - E\right)\Psi - \frac{\hbar^2}{2m}\left(2\tau\cdot\nabla + \nabla\tau\right)\Psi = 0 \qquad (9.66)$$

Equation (9.66) can also be written in a more compact way [28, 31]:

$$-\frac{\hbar^2}{2m}\left(\nabla + \tau\right)^2\Psi + \left(u - E\right)\Psi = 0 \qquad (9.67)$$

Equation (9.67) is the nuclear Schrödinger equation within the adiabatic framework for a group of states that forms a Hilbert space. In this sense equation (9.67) differs from equation (9.65) which is valid for any group of states. We notice that the non adiabatic coupling term appears in the same place in the Schrödinger equationwhere one would expect a vector potential, hence it is justified to consider it as a vector potential. This vector potential is a vector of matrices which do not necessarily commute, thus it is a non-Abelian vector potential. This of course raised the question if this non-Abelian vector potential appearing in the nuclear Schrödinger equation has something to do with non-Abelian vector potentials appearing in elementary particle theories such as the standard model, this will be the topic of concluding section.

9.5 Singularities

9.5.1 A Solenoid as a Model for the Seam

In previous sections we showed that the adiabatic Born-Oppenheimer eigenfunctions that form a Hilbert space, produce Curl-Divergence equations (see equation (9.27) and equation (9.38), respectively) which resemble the Curl-Divergence equations encountered in electrodynamics (in what follows we refer to these equations as C-D equations).

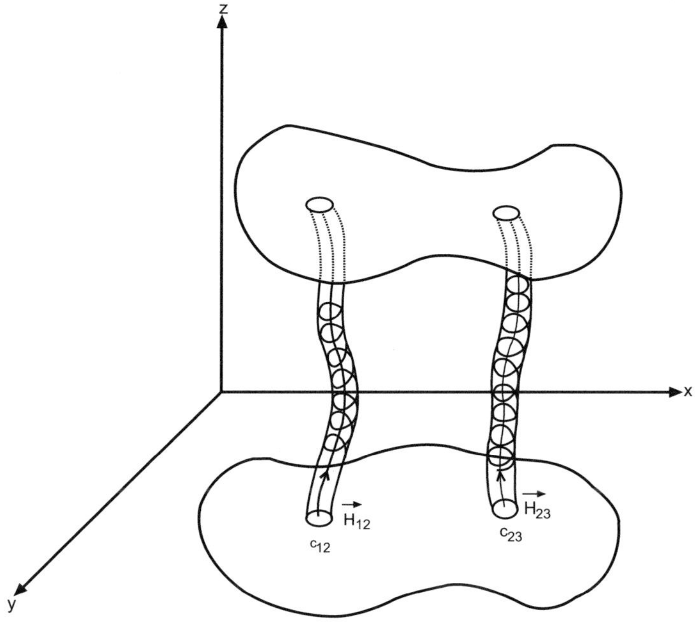

Figure 9.1: Simulation of a seam as a helical (electric) wire characteristic for a solenoid. Two seams are shown: one formed by a (1,2) ci and one by a (2,3) ci.

Based on this feature we present a new approach to treat the NACTs namely, to consider their spatial distribution as *vector potential* where at least some of them are produced by source-points reminiscent of electro-magnetic fields formed by charged particles (electrons, protons etc.). It will be shown that the source points in our case are the *ci* points - the points at which certain NACTs become singular (see Section 5.1 of [15]). Since *ci*s arrange themselves along (semi) *infinite* long contour lines – *seams* [32, 33, 34] – and since electric charges do not exist formally within the Born-Oppenheimer framework, the appearance of charges and field sources in the theory is striking. This analogy almost speaks for itself because a *seam* is reminiscent of the infinite (or semi-infinite) long and infinitesimal narrow *solenoid* which forms a magnetic field, $\vec{H}$ (see figure 9.1). To a good approximation it is known that solenoids produce magnetic fields only inside the solenoid, whereas outside the solenoid the field is zero, this should be expected for a one dimensional Hilbert subspace in which the field should vanish everywhere except for a thin tube containing the one dimensional singularity in which the vector potential is not analytic. In electrodynamics, the magnetic field $\vec{H}$ and the vector potential, $\vec{A}$ are connected according to the relation:

$$\vec{H} = Curl\vec{A} \tag{9.68}$$

Since the direction of $\vec{H}$ in the solenoid is along the solenoid's axis, equation (9.68) implies that $\vec{A}$ can be taken to be orthogonal to $\vec{H}$ and therefore also to the solenoid's axis. Thus, if for a given value of z the $x - y$ plane is chosen in such a way that the axis of the solenoid is perpendicular to it (and therefore is along the z-axis), then $\vec{H}$ possesses only a z-component and $\vec{A}$ only the (non-zero) components A_x and A_y. Assuming that the $\vec{A}$-solenoid system simulates correctly the τ - seam system this implies that if the seam is in the z-direction then $\tau_z \sim 0$ and τ_x and τ_y differ from zero.

There are, however, two major differences between the solenoid-$\vec{A}$ and the seam -$\vec{\tau}$ systems which cannot be ignored.

1. The vector potential $\vec{A}$ is Abelian, namely, made up of vector functions whereas $\vec{\tau}$ is usually a non-Abelian magnitude made up of (vector) *matrices*. Because of this difference the simulation may fail. However in Corollary at the end of Section 9.2.3 it was shown that the 2×2 $\vec{\tau}$ - matrices behave like Abelian and therefore a simulation, can be formed for a two-state Hilbert subspace. Moreover, the two-state case is exceptionally important because, usually, any non-Abelian $\vec{\tau}$ - matrix reduces, at the vicinity of any ci, to a 2×2 matrix and therefore becomes Abelian so that the simulation is always valid close enough to a given ci.(We remind the reader that as the distance from the given ci increases, the $\vec{\tau}$ - matrix, due to interactions with other ci's, expands to become N-dimensional, where $N > 2$, and therefore is non-Abelian.)

2. The $\vec{A}$-vector produced by a solenoid possesses a cylindrical symmetry whereas the spatial distribution of a NACT (even formed by a *single* ci) is rarely circular. For instance in case of a three-atomic systems it is frequently elliptic [35] (see section 3.2.3 of [15]). This difference is a source for difficulties because for a solenoid, $\vec{A}$ can be chosen so that it fulfills the equation $\vec{\nabla} \cdot \vec{A} = 0$ [36] (by this we do not mean that $\vec{\nabla} \cdot \vec{A}$ *has* to be zero, but it can be chosen to be so by choosing the Coulomb gauge) whereas in our case we usually have $\vec{\nabla} \cdot \vec{\tau} \neq 0$. It is important to emphasize that for molecular systems $\vec{\nabla} \cdot \vec{\tau}$ *has* to differ from zero.

In the present Chapter the molecular C-D equations are reconsidered for a different reason, namely, to prove that indeed the various elements of the $\vec{\tau}$ -matrix behave like (molecular) fields [37, 38, 39]. It will be shown that part of them have sources just like in electrodynamics and others, that lack sources, are induced by the fields with sources - a situation not encountered in electrodynamics but is encountered in non-Abelian gauge fields such as the gluons that appear chromodynamics. Both types will be studied employing the ordinary mathematical tools as applied within field theory.

In the next section we show, step by step, to what extent the two-state NACT behaves like an Abelian system.

9.6 Two-State (Abelian) System

9.6.1 The Non-Adiabatic Coupling Term as a Vector Potential

We start by stating the following Lemma:

Lemma

Given a two-state Hilbert subspace the corresponding two coupled adiabatic Schrödinger equations in equation (9.67) can be decoupled, for low enough energy, so that one equation is for a function $\phi(\vec{s})$ and the second for its complex conjugate, namely, $\phi^*(\vec{s})$. [40]

Proof

Taking advantage of equation (9.67) and equation (9.29) we replace, Ψ is replaced by Φ following the (constant) unitary transformation, thus:

$$\Psi = G\Phi \tag{9.69}$$

where G is chosen in such a way that it diagonalizes the τ-matrix [40, 41]:

$$G = \frac{1}{\sqrt{2}}\begin{pmatrix} 1 & 1 \\ i & -i \end{pmatrix} \tag{9.70}$$

Substituting equation (9.69), equation (9.70) and equation (9.29) in equation (9.67) (see above) yields the two coupled equations:

$$\begin{array}{l} -\frac{\hbar^2}{2m}\left(\nabla + i\tau\right)^2\phi_1 + \left(\frac{1}{2}(u_1+u_2) - E\right)\phi_1 + \frac{1}{2}(u_1-u_2)\phi_2 = 0 \\ -\frac{\hbar^2}{2m}\left(\nabla - i\tau\right)^2\phi_2 + \left(\frac{1}{2}(u_1+u_2) - E\right)\phi_2 + \frac{1}{2}(u_1-u_2)\phi_1 = 0 \end{array} \tag{9.71}$$

where τ_{12} is replaced by τ to simplify the notation. These two equations can be rearranged to become:

$$\begin{array}{l} -\frac{\hbar^2}{2m}\left(\nabla + i\tau\right)^2 \phi_1 + \left(u_1 - E\right)\phi_1 + i\frac{1}{\sqrt{2}}(u_2 - u_1)\psi_2 = 0 \\ -\frac{\hbar^2}{2m}\left(\nabla - i\tau\right)^2 \phi_2 + \left(u_1 - E\right)\phi_2 - i\frac{1}{\sqrt{2}}(u_2 - u_1)\psi_2 = 0 \end{array} \qquad (9.72)$$

where ψ_2, the (nuclear) adiabatic function associated with the upper state u_2, is put back again to replaced the difference $(\phi_1 - \phi_2) = -i\sqrt{2}\psi_2$ (employing equation (9.69) and equation (9.70)). Next we assume the energy to be low enough so that u_2, is classically forbidden and therefore $|\psi_2| \sim 0$ at every point. Consequently the term that contains ψ_2 can be neglected so that the two coupled equations become decoupled [40, 41]:

$$-\frac{\hbar^2}{2m}\left(\nabla \pm i\tau\right)^2 \phi + \left(u_1 - E\right)\phi = 0 \qquad (9.73)$$

where the subscripts 1 and 2 become redundant and are deleted. Thus we managed to derive a single (decoupled) Schrödinger equation which keeps its NACT contrary to the ordinary BO approximation [1, 2, 3]. It is important to mention that the derivation of a decoupled single Schrödinger equation from a, general, N-state system is discussed in Chapter eight of [15].

The outcome of this proof is that the NACT appears in the single-state Schrödinger equation in the same way as the magnetic vector potential appears in a Hamiltonian that describes the interaction of a charged particle with an external magnetic field [42, 43].

Comment: It is important to remind the reader that writing the Hamiltonian as in equation (9.73) guarantees, classically, that the magnetic force that acts on a charged particle is the Lorentz force [43], namely:

$$\vec{F} = \frac{e}{mc}\left[\vec{p} \times \vec{H}\right]$$

However due to equation (9.31) (see, also, next section equation (9.75)) $\vec{H} = 0$ and consequently the Lorentz force for the charged particles within the molecule (at least in close vicinity of the ci) is identically zero.

9.6.2 Two-State curl Equation

The two-state curl equation is introduced in section 9.2.2, in particularly through equation (9.30) and equation (9.31), where, it is presented, in terms of Cartesian Coordinates. Here and in what follows we are interested in a planar system described by two polar coordinates (φ, q), where the two are related to the Cartesian coordinates as:

$$\varphi = \tan^{-1}(y/x) \; ; \qquad q = \sqrt{x^2 + y^2} \tag{9.74}$$

Transforming from the Cartesian to the polar coordinates yields for the curl equation in polar coordinates (see equation (9.28) - equation (9.31)):

$$F_{q\varphi} = \frac{1}{q}\left(\frac{\partial \tau_\varphi}{\partial q} - \frac{\partial \tau_q}{\partial \varphi}\right) = 0 \quad \Rightarrow \quad \frac{\partial \tau_\varphi}{\partial q} - \frac{\partial \tau_q}{\partial \varphi} = 0 \tag{9.75}$$

where:

$$\tau_\lambda (\equiv \tau_{12\lambda}) = \left\langle \zeta_1 \,\middle|\, \frac{\partial}{\partial \lambda} \zeta_2 \right\rangle; \qquad \lambda = \varphi, q \tag{9.76}$$

Next we prove the following Lemma:

Lemma:

For a two-state Hilbert subspace that its line integral in equation(3.12) of [15] is quantized the radial component of $\vec{\tau}$, namely, τ_q, has to be single-valued [44].

Proof:

Integrating equation (9.75) with respect to both, q and φ, yields the result:

$$\begin{array}{l} \int_0^{2\pi} \tau_\varphi(\varphi, q_2) d\varphi - \int_0^{2\pi} \tau_\varphi(\varphi, q_1) d\varphi - \\ \int_{q_1}^{q_2} dq \left(\tau_q(\varphi = 2\pi, q) - \tau_q(\varphi = 0, q)\right) = 0 \end{array} \tag{9.77}$$

The quantization implies that equation(3.12) of [15] holds for every radius q so that we have:

$$\int_0^{2\pi} \tau_\varphi(\varphi, q_2) d\varphi = \int_0^{2\pi} \tau_\varphi(\varphi, q_1) d\varphi \quad (= n\pi) \tag{9.78}$$

Returning to equation (9.77), we get for $\tau_q(\varphi, q)$:

$$\int_{q_1}^{q_2} dq \left(\tau_q(\varphi = 2\pi, q) - \tau_q(\varphi = 0, q)\right) = 0 \tag{9.79}$$

Thus, $\tau_q(\varphi, q)$ is a single-valued function with respect to φ for any q.

9.6.3 The (Extended) Stokes Theorem

Stokes theorem implies:

$$\oint_{\Gamma} d\vec{s} \cdot \vec{\tau}(\vec{s}) = \int\int_{\sigma} F d\sigma \tag{9.80}$$

where F is introduced in equation (9.75), τ is the $(1, 2)$ NACT vector (not a matrix) mentioned earlier and the closed contour Γ is the border line of the assigned planar region σ. On the l.h.s. of equation (9.80) we encounter a *line* integral and on the r.h.s. a *surface* integral.

It seems as if equation (9.80) is not fulfilled within the present formalism because the expression on the r.h.s. is always zero (see equation (9.75)) whereas the expression on the l.h.s., due to the quantization condition (see equation(3.12) of [15]), usually differs from zero. The reason for this discrepancy has to do with the integration over σ which cannot be correctly carried out because F is not defined at the singular *ci*-points (these points were defined in section 1.3.2.2 of [15] as *pathological* points). To correct for this defect, we recall that the *ci*-points form a *finite* set of (isolated) points and therefore at each such a point the (φ, q) tensorial component of F, namely, $F_{q\varphi}$ has to be infinitely large to ensure that the surface integral in equation (9.80) yields a *non-zero* contribution. To continue we consider the case of one single *ci*. In order to guarantee the quantization in equation(3.12) of [15]), we assume for $F_{q\varphi}$, at the *ci*-point, the following expression [45]:

$$F_{q\varphi} = 2\pi f(\varphi, q) \frac{\delta(q)}{q} \tag{9.81}$$

Here $\delta(q)$ is the Dirac's Delta function and $f(\varphi, q)$ is an analytic, but yet undetermined, function.

To continue we concentrate on a circular region, that contains one *ci* at $q = 0$ and unite equation (9.81) and equation (9.75) to form an extended version for the curl equation, namely:

$$F_{\varphi q} = \frac{1}{q}\left(\frac{\partial \tau_{\varphi}}{\partial q} - \frac{\partial \tau_{q}}{\partial \varphi}\right) - 2\pi f(\varphi, q)\frac{\delta(q)}{q} = 0 \tag{9.82}$$

Next, considering again equation (9.80) and employing equation (9.82) for $F_{q\varphi}$ we obtain that the Stokes Theorem leads to the following expression [10, 46]:

$$\oint_{\Gamma} \frac{1}{q}\tau_{\varphi}(\varphi, q) q d\varphi = 2\pi \oint_{\sigma} \frac{1}{q} f(\varphi, q)\delta(q) q d\varphi dq \tag{9.83}$$

In writing equation (9.83) it is assumed that equation (9.75) is fulfilled at every point in σ (including $q = 0$).

To continue we perform the integrations on both sides of equation (9.83) and get that in order to guarantee the equality the function $f(\varphi, q = 0)$ has to fulfill the condition:

$$\int_0^{2\pi} \tau_\varphi(\varphi, q) d\varphi = \pi \int_0^{2\pi} f(\varphi, q = 0) d\varphi \tag{9.84}$$

However recalling equation (9.78) we obtain that $f(\varphi)\{\equiv f(\varphi, q = 0)\}$, just like $\tau_\varphi(\varphi, q)$, has to be quantized, namely:

$$\int_0^{2\pi} f(\varphi) d\varphi = n \tag{9.85}$$

where n is an integer.

We continue to elaborate on equation (9.82). This equation is defined for both $q = 0$ and $q > 0$ and therefore can be considered as the extension to equation (9.75) [45]:

$$\frac{\partial \tau_\varphi}{\partial q} - \frac{\partial \tau_q}{\partial \varphi} = 2\pi f(\varphi) \delta(q) \tag{9.86}$$

It is noticed that adding the r.h.s in equation (9.86) is, in fact, a way to incorporate, in equation (9.75), the boundary condition, at $q = 0$. The solution of equation (9.86) (for $q > 0$) is given in the form:

$$\tau_\varphi(q,\varphi) - \int_\varepsilon^q dq \frac{\partial \tau_q}{\partial \varphi} = \pi h(q) f(\varphi) \tag{9.87}$$

where $\epsilon \longrightarrow 0$ and $h(q)$ is the Heaviside step function:

$$h(q) = \begin{cases} 1 & q > 0 \\ 0 & q < 0 \end{cases} \tag{9.88}$$

The fact that equation (9.87) is a solution of equation (9.86) can be verified by substitution. Since q is a radius it is always positive and therefore equation (9.87) can be written, without loss of generality, as:

$$\tau_\varphi(q,\varphi) - \int_\varepsilon^q dq \frac{\partial \tau_q}{\partial \varphi} = \pi f(\varphi) \tag{9.89}$$

Next, assuming $\tau_q(\varphi, q)$ to be *finite* at the vicinity of $q \sim 0$, it is noticed that the second term on the l.h.s. of equation (9.89) for $q \rightarrow \epsilon(\sim 0)$ can be ignored so that we find [45]:

$$f(\varphi) = \frac{1}{\pi}\tau_\varphi(q \sim 0,\varphi) \tag{9.90}$$

This result implies that the angular dependence of the curl term in equation (9.81) is identical (up to a division by π) to the *angular* component of $\vec{\tau}$ as calculated in the close vicinity of the *ci* (i.e. $q \sim 0$). In what follows the angular component at $q \sim 0$ is termed as the *virgin* function – virgin in the sense that its values are not damaged by NACTs of other *ci*'s. The virgin functions will be used as boundary conditions to solve the C-D equations.

An unavoidable difficulty of the present theory is that the radial component of $\vec{\tau}$, namely, $\tau_q(\varphi, q)$, remains, in fact, unknown. Moreover except for it being single-valued in configuration space we are not aware of any special features. We *assumed* it to be finite for $q \sim 0$ but no clue is given with regard to its values for $q > 0$. In fact distinguishing, for $\tau_q(\varphi, q \sim 0)$, between being finite or identically zero is not essential because the angular component of $\vec{\tau}$, namely, $\tau_\varphi(\varphi, q)/q$ becomes infinitely large and therefore completely dominates $\vec{\tau}$, in this region. Since at the *ci*-point the values of $\tau_q(\varphi, q)$ are of minor importance we continue to ignore it at the whole region of interest. In other words we *assume* that for an isolated *ci*, $\tau_q(\varphi, q) \sim 0$ for any q-value. We are aware that this is a far reaching *assumption* and therefore, this assumption was under a detailed scrutiny in numerical treatments [47, 48, 38].

9.7 Conclusion

There are a couple of major differences between molecular and elementary particle non-Abelian fields:

1. The typical energies in molecular systems are much lower than the energies involved in elementary particle physics.

2. Elementary particle gauge theories are Lorentz invariant and the vector potentials are four vectors (of matrices), in molecular physics the dimension of the vector potential depends on the number of nuclear degrees of freedom, for a single nucleus the number is three. This fact is of course not surprising since the molecular vector potentials are derived from a theory which is not Lorentz invariant.

3. The dimensions of the matrices in the molecular vector potentials is the number of coupled electron states (which is equal to the dimension of the sub-Hilbert space). In elementary particle "standard model" the dimension of the vector potential matrices is arbitrary and is chosen in such a way as to fit experimental data. The dimensions of the matrices which belong to $SU(3) \times SU(2) \times U(1)$ cannot be explained even by string theory which allows many possible models.

4. The non-Abelian field F_{ij} that appears in molecular physics vanishes for a complete Hilbert space. The non-Abelian field does not vanish in the context of elementary particle physics and in fact provides the dynamics necessary for the field propagation. This again can be attributed to the fact that molecular non-Abelian field theory is developed in a non-relativistic context that allows instantaneous action at a distance. We notice, however, that the molecular non Abelian field will not vanish when the sub-Hilbert space is not decoupled effectively from the rest of electron Hilbert space. Moreover, the field will not vanish in the case of singularities (conical intersections) that appear when energy levels cross each other.

Having said that, it would be beneficial if there could be found a connection between molecular non-Abelian fields and elementary particle non-Abelian fields which is not a strictly formal one. The reason is that if such a connection could be found it can shed a light on the "standard model" which has a remarkable experimental success, but its structure (and numerous constants) can not be explained or justified. In contrast the structure of the molecular non-Abelian fields can be uniquely determined based on the standard Schrödinger equation and the system under study. If the sought after connection could be found it may give us hope to understand the "standard model" and its success. In order to achieve this an extension of molecular dynamics and the Born-Oppenheimer approach to the relativistic regime is paramount. If this can be done perhaps an advancement can be achieved in the field theory of non-Abelian gauge fields which will reduce the number of assumptions needed to formulate such a theory. The future certainly holds for us more advances in the classical theory of fields.

Bibliography

[1] M. Born and J.R. Oppenheimer, "Zur Quantentheorie der Molekeln", *Ann. Phys.(Leipzig)*, vol. 84, 457, 1927.

[2] M. Born, *Festschrift Gott. Nach. Mathem. physik. Klasse*, 1, 1951.

[3] M. Born and K. Huang, Dynamical Theory of Crystal Lattices. Oxford University Press: New York, 1954. Chap. IV.

[4] P.A.M. Dirac, The Principles of Quantum Mechanics. Oxford University Press: New York, 1958. Chap. II.

[5] A. Messiah, Quantum Mechanics. North Holland Publishing Company: Amsterdam, 1964, Vol. I, section V.2.

[6] R. Baer, "Born Oppenheimer Invariants Along Nuclear Paths," *J. Chem. Phys.*, 117, 7405, 2002.

[7] M. Guidry, Gauge Field Theories an Introduction with Applications. Wiley Science Paperback Series: New York, 1999. (a) Chapter 13 (b) Chapter 7.

[8] M.E. Peskin and D.V. Schroeder, An Introduction to Quantum Field Theory. Perseus Books: Cambridge, Massachusetts, 1995. Section 15.2.

[9] C.N. Yang, and R.L. Mills, "Conservation of Isotopic Spin and Isotopic Gauge Invariance," *Phys. Rev.*, vol. 96, 191, 1954.

[10] R. Englman and A. Yahalom, "Signed phases and fields associated with degeneracies," *Acta Univ. Debreceniensis Series Physica et Chimica*, vol. 283, 34-35, 2002.

[11] R. Englman, and A. Yahalom, "Complex States of Simple Molecular Systems," In: The Role of Degenerate States in Chemistry, editors M. Baer, and G.D. Billing, Adv. Chem. Phys., vol. 124, 197 section VI B, John Wiley and Sons: New York, 2002.

[12] R. Baer, D.J. Kouri, M. Baer and D.K. Hoffman, "General BornOppenheimerHuang approach to systems of electrons and nuclei," *J. Chem. Phys.*, vol. 119, 6998, 2003.

[13] M. Baer, "Adiabatic and diabatic representations for atom-molecule collisions: Treatment of the collinear arrangement," *Chem. Phys. Lett.*, vol. 35, 112, 1975.

[14] M. Baer, "Introduction to the theory of electronic non-adiabatic coupling terms in molecular systems ," *Phys. Rep.*, vol. 358, 75. 2002.

[15] M. Baer, Beyond Born-Oppenheimer, Conical Intersections and Electronic Nonadiabatic Coupling Terms. John Wiley and Sons: New-York 2006.

[16] W.D. Hobey, and A.D. McLachlan, "Dynamical Jahn-Teller Effect in Hydrocarbon Radicals," *J. Chem. Phys.*, vol. 33, 1695, 1960.

[17] J.D.Jackson, Classical Electrodynamics, Third Edition. Wiley: New York, 1999. Chapter 6.

[18] L.D. Landau and E.M. Lifshitz, The Classical Theory of Fields. Addison-Wesley Press, Inc.: Cambridge, Mass, 1951. p. 47.

[19] M. Baer, "Classification of topological effects in molecular systems," *Chem. Phys. Lett.*, vol. 322, 520, 2000.

[20] M. Baer, T. Vertesi, G.J. Halasz, A. Vibok, and S. Suhai, "Beyond BornOppenheimer," *Faraday Disc.*, vol. 127, 337, 2004.

[21] H. Feshbach, "Unified theory of nuclear reactions," *Ann. Phys. (N.Y.)*, vol. 5, 357, 1958.

[22] C.L. Shoemaker, and R.E. Wyatt, "Feshbach Resonances in Chemical Reactions," *Adv. Quant. Chem.*, vol. 14, 169, 1981.

[23] M. Baer, and G.D. Billing Editors, The Role of Degenerate States in Chemistry. Adv. Chem. Phys., vol. 124, John Wiley and Sons: New York, 2002.

[24] M. Baer and R. Englman, "Electronic non-adiabatic transitions: the line integral and approximations," *Chem. Phys. Lett.*,vol. 335, 85, 2001.

[25] A.D. McLachlan, "The wave functions of electronically degenerate states," *Mol. Phys.*, vol. 4, 417, 1961.

[26] M. Baer, "Topological effects in molecular systems: an attempt towards a complete theory," *Chem. Phys.*, vol. 259, 123, 2000. see Appendix B.

[27] M. Baer, "Electronic non-adiabatic transitions derivation of the general adiabatic-diabatic transformation matrix," *Molec. Phys.*, vol. 40, 1011, 1980.

[28] T. Pacher, L.S. Cederbaum, and H. Köppel, "Adiabatic and Quasidiabatic States in a Gauge Theoretical Framework," *Adv. Chem. Phys.*, vol. 84, 293, 1993.

[29] M. Baer, "Existence of Diabatic Potentials and the Quantization of the Nonadiabatic Matrix," *J. Phys Chem. A*, vol. 104, 3181, 2000.

[30] H.C. Longuet-Higgins, "Some Recent Developments in the Theory of Molecular Energy Levels," In: Advances in spectroscopy. Vol. II. H.W. Thompson ed. Interscience: New York, 429, 1961.

[31] F.T. Smith, "Diabatic and Adiabatic Representations for Atomic Collision Problems," *Phys. Rev.*, vol. 179, 111, 1969.

[32] E.E. Nikitin, "Theory of nonadiabatic transitions. Recent development of Landau- Zener model," In: Chemische Elementarprozesse. Berlin: Spriger Verlag, 1968, pp. 43-77.

[33] Z.H. Top, and M. Baer, "An exact quantum mechanical treatment of charge transfer at low energies for the H++H2 reactive collinear system," *J. Chem. Phys.*, vol. 64, 3078, 1976.

[34] T. Vertesi, and E. Bene, "Pair annihilation of conical intersections and a study to infer the phenomenon," *Chem. Phys. Lett.*, vol. 392, 17, 2004.

[35] M. Baer, A.M. Mebel, and R. Englman, "Conical intersection revisited: extension to an elliptic form," *Chem. Phys. Lett.*, vol. 354, 243, 2002.

[36] R.P. Feynman, R.B Leighton and M. Sands, The Feynman Lectures on Physics. Addison-Wesley Publishing Company: New York, 1964. Vol. II: (a) Sect. 14.1; (b) Sect. 14.4.

[37] M. Baer, "Gauge transformation to study electronic non-adiabatic coupling terms in molecular systems," *Chem. Phys. Lett.*, vol. 360, 243, 2002.

[38] T. Vertesi, A. Vibok, G.J. Halasz, and M. Baer, "A field theoretical approach to calculate electronic Born-Oppenheimer coupling terms," *J. Chem. Phys.*, vol. 121, 4000, 2004.

[39] A. Vibok, G.J. Halasz, A.M. Mebel, S. Hu, and M. Baer, "An analytic-numeric approach to calculate electronic non-adiabatic coupling terms: study of the C2H molecule and the H2+H system," *Int. J. Quant. Chem.*, vol. 99, 594, 2004.

[40] M. Baer, and R. Englman, "A modified Born-Oppenheimer equation: application to conical intersections and other types of singularities," *Chem. Phys. Lett.*, vol. 265, 105, 1997.

[41] M. Baer, S.H. Lin, A. Alijah, S. Adhikari, and G.D. Billing, "Extended approximated Born-Oppenheimer equation. I. Theory," *Phys. Rev. A*, vol. 62, 032506, 2000.

[42] P.A.M. Dirac, "Quantised Singularities in the Electromagnetic Field," *Proc. R. Soc. London A*, vol. 133, 60, 1931.

[43] H. Goldstein, Classical Mechanics. Addison-Wesley Publishing Company Inc.: Cambridge, Mass., 1956. Section 1-5.

[44] M. Baer, A.M. Mebel and G.D. Billing, "Curl equations as substratum for the derivation of the electronic nonadiabatic coupling terms," *Int. J. Quant. Chem.*, vol. 90, 1577, 2002.

[45] M. Baer, "Curl condition revisited," *Chem. Phys. Lett.*, vol. 349, 84, 2001.

[46] T. Vertesi, A. Vibok, G.J. Halasz, A. Yahalom, R. Englman and M. Baer, "The Electronic Non-Adiabatic Coupling Matrix: A Numerical Study of the Curl Condition and the Quantization Condition Employing the Mathieu Equation," *J. Phys. Chem. A*, vol. 107, 7189, 2003.

[47] T. Vertesi, A. Vibok, G.J. Halasz, and M. Baer, "Derivation of the electronic nonadiabatic coupling field in molecular systems: An algebraic-vectorial approach," *J. Chem. Phys.*, vol. 120, 8420, 2004.

[48] A. Vibok, T. Vertesi, E. Bene, G.J. Halasz, and M. Baer, "Vector-Algebra Approach To Obtain Molecular Fields from Conical Intersections: Numerical Applications to H + H2 and Na + H2," *J. Phys. Chem. A*, vol. 108, 8590, 2004.

CHAPTER 10

Collapse of Wave-Packet component Phases: A Dispersion Relation Treatment

Robert Englman[1]

Abstract

Kramers-Kronig or dispersion relations that arise from the analytic behavior of electromagnetic and other classical waves are here adapted to quantum waves. In the framework of theories that view the wave packet (wp) collapse as a time-continuous process, we postulate analytic properties for the component-amplitudes in a wp as functions of a complex time. We then construct a model which embodies the removal of the non-selected components in the aftermath of the measurement, to be ultimately followed by a thermal-equilibrium like superposition state of the system. Conjugate relations hold between component-moduli and phases and these show that a *non-selected* component (one that vanishes in the measurement) acquires within the duration of the collapse a fast oscillating phase factor. Thus, by virtue of mathematical properties, both phase-decoherence and amplitude-decay have to occur in a collapse process, indifferently to the physical mechanism.

[1]Department of Physics and Applied Mathematics, Soreq NRC, Yavne 81800, Israel
e-mail: englman@vms.huji.ac.il;

10.1 Introduction

As mentioned before the theory of quantum mechanics is mathematically speaking a theory of a complex field that must satisfy a complex partial differential equation, the Schrödinger equation. This field is unique in the sense that its modulus square represents a probability. However, what is probable before measurement occurs, has taken place after measurement was performed, hence the wave function must collapse. Unfortunately Schrödinger equation in its standard form does not describe collapse phenomena and is restricted solely to unitary development. In this chapter we will try to describe mathematically what happens to a wave function during collapse, in particular the behavior of the wave function phase will be studied. This is considered an advancement in the field theory of quantum mechanics since it widens our understanding of the wave function behavior even when Schrödinger equation can not be easily applied.

10.1.1 Historical background

The relation between phases and magnitudes of classical wave-fields was the subject of consideration by Lord Rayleigh [1], who remarked that through interference between the probing and a probed wave the magnitude and phase of acoustic waves can each be determined, for example, by finding surfaces of minimum and zero magnitudes. In optics, in the context of the "complex analytic signal" (an electromagnetic field with only positive frequency components), the positions of field-zeros (from which the phase can be determined) and the intensity represent two sets of interdependent information, that are intertwined by the analytic property of the wave [2]. Here we adapt essentially the same mathematical apparatus (dispersion relations based on complex analysis) to investigate quantum mechanical collapse. The possibility (and presumed success) of such adaptation is founded on the famous words of Wigner [3]:

"The enormous usefulness of mathematics in the natural sciences is something bordering on the mysterious and there is no rational explanation for ... this uncanny usefulness of mathematical concepts...

The complex numbers provide a particularly striking example of the foregoing. Certainly, nothing in our experience suggests the introducing of these quantities... Let us not forget that the Hilbert space of quantum mechanics is the complex Hilbert space with a Hermitian scalar product. Surely to the unpreoccupied mind, complex numbers ... cannot be suggested by physical observations. Furthermore, the use of complex numbers is not a calculational trick of applied mathematics, but comes close to being a necessity in the formulation of the laws of quantum mechanics. Finally, it now [1972] begins to appear that not only complex numbers but analytic functions are destined to play a decisive role in the formulation of quantum theory. I am referring to the rapidly developing theory of dispersion relations. It is difficult to avoid the impression that a simple miracle confronts us here [i.e. in the agreement between the properties of the hypernumber $\sqrt{(-1)}$ and those of the natural world]."

10.1.2 Collapse interpretations

Collapse of a wave function has been defined (with the distinctive adjective of "genuine") as the "disappearance of all but one component [of a superposition or of a wave-packet (wp)] from reality" [4, 5]. In an alternative, ensemble description of the measurement process, too, the collapse stage is brought about by the "disappearance of all but one" matrix elements (all the off-diagonal ones) in each row of the density matrix of the observed system. An article in [4] reviews several modes of collapse [6]. A further process is outlined in [7]. (In the penultimate section, we shall document from David Bohm's writings [8] the commonness of his approach with our analysis and will also address there the aforementioned alternative descriptions of collapse, namely, as given in terms of component-amplitudes or density-matrices. Until then we will keep our attention on the unifying "disappearance" aspect in theories of collapse. Also, throughout, we shall restrict our attention to the mathematical consequences of the collapse, irrespective of the physical mechanism controlling it. We are not proposing any physical mechanism.)

According to the orthodox view, a measurement in quantum mechanics ends with the collapse. After the collapse everything starts anew [9]. Opposing this view are a number of other approaches (including several recent ones), according to which the state of the measured system evolves on some minute time-scale continuously even during a collapse, though at that juncture the evolution may appear as undergoing an abrupt jump when observed on an ordinary time-scale. The difference in the two time-scales arises from the macroscopic nature of the apparatus (and the environment), though other effects (like gravity) may also play a role. Within the latter theories, there still remains a further point of difference as to whether the collapse is due to phase-decoherence (as e.g., in [10] -[12]), or is due to the decay of the magnitude of some quantity without any remarkable change in the phase attached to that quantity [13].

We shall show below that the these two pathways of collapse merge into a single one, in the sense that the vanishing of component modulus in a wave function entails a cataclysmic change in the associated phase, as well as *vice versa*. To demonstrate this, we construct in this chapter a mathematical model for a continuouscollapse!continuous collapse. The model requires some limiting procedures. However, the elaborate and formal details in this work should not hide the fact that the following elementary considerations make the claim of this chapter practically self-evident:

It is known that zeros (or nodes) in wave functions almost invariably increase (numerically) the phase. Examples of this are the textbook result linking the energy (=the time derivative of the phase) of the electronic states in a hydrogen atom or of a harmonic oscillator to the number of nodal surfaces or, in chemical physics, the excess energy of anti-bonding molecular orbitals, or the π Berry-phase tied to a zero in the component amplitude [14], or the Morse-Maslov index that increases whenever the trajectory meets a caustic (which is another example of singularity in the logarithm of the wave-function)[15]. It is therefore expected that the vanishing of a non-selected wave-function component in a measurement should also enhance the corresponding phase. However, since the vanishing is not just at a point or on a nodal plane in a multi-dimensional coordinate space, but over a continuum in time, the enhancement of the phase takes a complicated form. This aspect brings in an unusual element, so we have found it necessary to build a detailed model which involves some limiting procedures. Yet the basics of the effect seem quite elementary (though not so far fully recognized).

The other elementary point worth *re*-stressing is the logical contingency of the present chapter to *continuous* theories of collapse. Only within the framework of these theories do we make the claim of this chapter. Here again, the mathematical implications of the continuous formulation of the measurement process have not been adequately covered so far.

10.2 After the collapse

Theories and discussions of the measurement process are so numerous as to defy an adequate referencing. What characterizes these studies is that they look at the instant preceding or coinciding with the measurement, but only rarely, if at all, consider what happens with the state component amplitudes after the measurement. Exceptions noted by us are [16]-[18], of which the first depicts the history of the state at and after collapse and the last shows a component-modulus before, during and after collapse.

10.2.1 Phase behavior in a pointer basis model

In Ref. [16] Zurek presented a formal description of a collapse that takes place in conjunction with the selection by the environment of a macroscopic coordinate monitoring the measurement. The key mathematical quantity in the analysis is the time (t) dependent "correlation amplitude" ($z_{ij}(t)$ in Eq. (3.7) of [16], or related quantities there in Eq. (4.5) and in Fig.2), that regulates the decay of the off-diagonal elements of the density-matrix. $z_{ij}(t)$ is formally given (when we drop the double indices) as the complex function

$$z(t) = \sum_k p_k e^{-i\omega_k t} \tag{10.1}$$

in which p_k are "effective weights" of environment states and the ω_k's are frequency differences. The source of both is the interaction with the environment and, since the latter possesses many uncontrollable factors, the frequencies are taken as random quantities in some range. Changing the chosen frequency range will only affect the time scale and is immaterial in the present context. However, equation (10.1) holds only for time that is zero (the beginning of the measurement process) or positive (subsequent times). We shall later argue that for a closed system, the definition must cover the whole time range between $-\infty$ and ∞, but in this section we analyse the model as presented in [16]. (Some difficulties in energy driven collapse relevant also to this model have recently been noted [17].)

The salient property of $z(t)$ is that it vanishes soon after the initiation at t=0 of the measurement (defined in [16] as the interaction of the apparatus with its macroscopic environment) and that it will recur to a finite value only very much later. "Soon" (or "the duration of the collapse process", to be denoted in this work by Δt_c) will depend on the frequency spread and (less critically) on the number of k-states in the above sum. The ultimate vanishing of the modulus $|z(t)|$ (the essential element in the treatment of [16]) is beyond doubt. We now study the phase of this, almost-periodic function

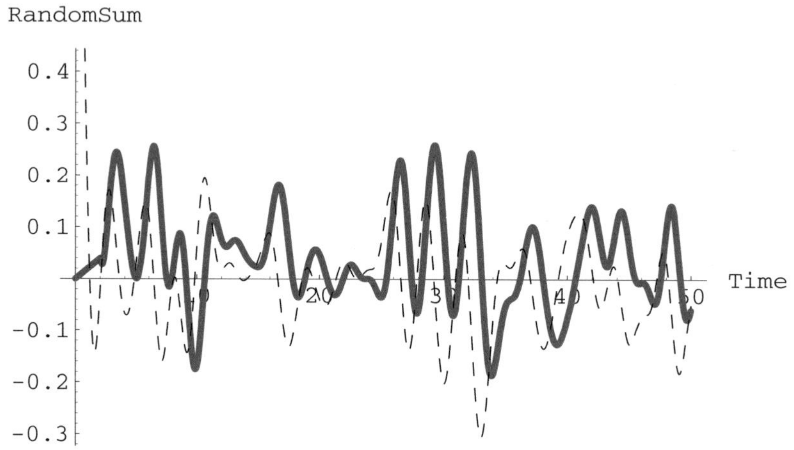

Figure 10.1: The sums in equation (10.2) against time [in units of $\frac{2\pi}{\omega_0}$, with ω_0 arbitrary]. The continuous curve shows the *sine* sum in equation (10.2) with all p_k equal to 1 and with the frequencies ω_k randomly distributed in the interval $(0, \omega_0)$. The broken curve is the *cosine* sum. For the most cases, the two curves alternate in crossing the abscissa.

$z(t)$. The phase is given by

$$\arg z(t) = -\arctan \frac{\sum p_k \sin \omega_k t}{\sum p_k \cos \omega_k t} \tag{10.2}$$

and its development depends on the succession of zeros in the numerator and the denominator. To see this, let us just think of $z(t)$ consisting of only the first terms in the sums and we immediately recognize that, whenever a zero in the numerator ($\sin \omega_1 t$)) is followed by a zero in the denominator ($\cos \omega_1 t$)) followed by a further zero in the numerator, there is a growth of phase of π. On the other hand, two successive numerator (or denominator) zeros return the phase towards its value before the zeros. We have computed the quantities in the numerator and in the denominator for the special and simple case that all p_k are equal and the ω's are randomly distributed between 0 and 1 (Figure 10.1).

It is evident from the two curves that the great majority of numerator- and denominator-zeros alternate. On the other hand, the distribution of zeros of either of these quantities has been studied by Kac [19] (recalled also in [16]) and was found (under conditions satisfied by the sums above) to vary asymptotically linearly with t.

[In fact, for uniform, constant p_k's and densely distributed, random frequencies in the range $(0, \omega_0)$, it is legitimate to approximate the sums in equation (10.2) by integrals over this range to obtain

$$\arg z(t) \approx -\arctan \frac{1-\cos \omega_0 t}{\sin \omega_0 t} = -\frac{\omega_0 t}{2} \tag{10.3}$$

This asymptotic value has been confirmed by us *on the average* by a numerical computation of the phase using the original expression in equation (10.2) .]

A technical detail in the computational procedure leads to a new, theoretically significant result, as follows: We have used Mathematica to evaluate equation (10.2) numerically. However, since Mathematica (in common with other standard algorithms) restricts angles to lie between $-\pi$ and π, it is not possible to obtain angles beyond these values. To establish the asymptotically monotonic variation of the angle with t, we have first *differentiated* the expression in equation (10.2) with respect to t, and then *integrated* the differential between zero and t. The outcome for the (absolute value of the) argument function is, as already noted, a linearly increasing function of t, barring some stochastic deviations due to random sampling errors.

We wish to note that the above procedure is more than a computational trick. Rather, it suggests an operational way to obtain phase differences between components in a quantum mechanical superposition, such that the phases are not restricted to a 2π range. To achieve this, all we need to do is to make a large number of observations (numbering, say, M x $T >> 1$, with M and T about to be described) on an ensemble of identically prepared superpositions. Suppose that M observations taken at a particular time instant t are sufficient to determine to any desired degree of accuracy the values of the component-moduli and the cosines of the component-phase differences. From the cosines we get the phase differences, subject to uncertainties of sign and shifts by arbitrary multiples of 2π in every phase-difference. When this procedure is repeated at a further large number T of subsequent, closely spaced instances and the *continuity of the phases as functions of time is assumed*, one obtains phase values that are not restricted to any finite range.

This establishes that component-phase-increments between different instances can be operationally determined to any arbitrary extent and not just between (0, 2π) (this means that "high values of a phase possess an ontology"). It remains an open question, whether the extra phases can ever show up in observation, as in expectation values (namely, "do they possess a phenomenology?"). However, the Friedel sum rule expresses the difference at different energies between *phases* of the eigenvalues of the scattering matrix in terms of the number of particles located in the energy range, and the latter number is clearly observable and not restricted to a small integer [20]. Some other aspects of the question are treated in [21].

Putting in numbers, if the frequency spread is over a range of 10^{36} inverse seconds (which is perhaps a fair number for the total energy of a small macroscopic device containing 10^{22} atoms, each being in an eigenstate of relative energy $10^{14}s^{-1}\hbar$), then after 10^{-20} seconds one gets a phase of 10^{16} radians. This represents a fast oscillatory behavior (around some mean quantity associated with a macroscopic device, such as the position of a pointer), which is impossible to distinguish from zero, even if the modulus of the "correlation amplitude" remained finite.

We thus find that in the "correlation amplitude" introduced in [16] there is an interplay of behavior between the vanishing modulus and the acquisition of a large phase. (But at very short times, which are mere fractions of ${\omega_0}^{-1}$, when the modulus is still finite, the phase is small, too.)

10.3 Continuous Theories with Analytic Properties

10.3.1 "Continuous" and "Theories"

By "continuous" we mean a continuity (on a suitably fine time-scale) of a time-dependent quantity that formally characterizes the system, and this throughout the time range between $t = -\infty$ and $t = \infty$.

"Theories" are theoretical proposals, like several ones mentioned in detail or in passing in [4], that formulate collapse as a consequence of a time dependent Schrödinger equation applied to a closed system, which includes (perhaps subject to small modifications) the microscopic system, the device and their surroundings. We do not intend to formulate here such a theory, but only to high-light mathematical consequences of the theories that be. Though these may differ importantly in detail and in physical contents, by our understanding they agree on viewing collapse as continuous.

Because the system is closed, its Hamiltonian is time independent. The time dependence enters through postulating very far in the past (at a time that will be denoted by $t = -\infty$) a superposition state, or a wave-packet (wp), whose component coefficients (amplitudes) evolve with time. Without loss of generality, we assume that the phase of each amplitude starts with a zero value. (Since we shall find that close to the time of collapse many of the phases will increase very fast, a small or even finite inaccuracy in the phases very far in the past will not affect our results.)

Lower half-plane analyticity

We now carry the requirement of continuity a step farther to make the major assumption of our treatment.

This is, that the component amplitudes in the wp of the system are not only continuous and well-defined functions throughout the range between $t = -\infty$ and $t = \infty$, but that they are analytic (regular) in the lower part of the complex time-plane $t = t' + it"$, including the whole real axis. (It is not easy to extend ally Zurek's model in the previous section, originally defined for non-negative times only, to negative times and still to represent a measurement with a proper time arrow.) The distinguishing feature of the *lower* half plane is that energies in a closed and bound system possess a lower bound [22]. The expectation of analyticity is justified by the fact that component amplitudes are solutions of time dependent differential equations (the Schrödinger equation). We argue that no matter how complicated the system (comprising the observed quantum system, the measuring device and its environment) the governing factor is a differential equation in time with non-divergent terms, for which some such analytic behavior is expected. In fact, wave packets of several sorts have the property of being analytic in the lower half of the complex t-plane. Free particle wp's [23], the Gaussian wp's of Heller [24], as well as other wp's studied by us in the past [6, 14, 25] have the stated property of analyticity. Naturally, for a complicated system one expects to find wp's that are sums and products (for several particles) of the elementary, "text-book" wp's (cf. [24]). Still these should not have singularities that are not already present in the elementary wp's. They may, however, pick up new zeros (modes) in the wp; these are of some importance to us and will be treated later. We re-emphasize that in closed systems, like the ones we are considering, time dependent component amplitudes can only arise from states that were initially built up as non-monoenergetic wp's.

In the previous paragraph we have made the time-analyticity assumption to apply to the time-dependent component-coefficients of a wp. However, this assumption can also be expected to hold for the elements a density matrix, which are also time dependent and are solutions of first order differential equations (in time). Further, both quantities share the property of collapse expressed at the beginning of this article, namely, that (many of) their members vanish as part of the collapse process. We return to this remark in the penultimate section of this work.

From now on, for simplicity, we shall restrict our treatment to a time-evolving wave packet (wp) or superposition, rather than to a density matrix (except for a short discussion at the end of section 10.5), expecting that the outcome for both will be similar. We shall write the wp as

$$\Psi(t) = \sum_n c_n(t)\psi_n \tag{10.4}$$

where n labels the components in the prepared superposition and ψ_n is one of the states of the quantum system that the measurement process can select (e.g., electronic spin states). When the label of one of these components is $S \quad (n = S)$, this means that this component is actually selected by one particular measurement, while all other components are not selected. The labels of the latter are generically designated NS. All amplitudes $c_n(t)$ are time dependent and complex.

Depending on the level of sophistication (and the extent of system - environment entanglement) that is desired, the amplitudes must be conceived as functions of a great many coordinates other than that of the quantum system ("Entangled system"). What is essential is that in the course of collapse the non-selected $c_{NS}(t)$ will fast convert from a finite value to a vanishingly small one, while the selected $c_S(t)$ will convert from a finite value to another finite value. This is our collapse model for a superposition state. Our aim now is to explore the consequences of this behavior on the phase of the coefficients.

10.3.2 Further assumption on the analytic behavior

Our other requirement is that every component-coefficient $c_n(t)$ varies fairly smoothly on a large semi-circle on the lower half of the complex plane. A better insight into this assumption will be gained, when the model is introduced in the next subsection, since in that model $|c_n(t)|$ has the same value at both extremities of the real axis (namely, the value appropriate to a "thermal equilibrium", to be explained there) and it may well be presumed that the coefficient will not change by much along the large arc that joins these extremities along the large semi-circle. The bases for this assumption are that, on one hand, it is hard to see any real physical effect propagating that far to the imaginary axis ($t" \to -\infty$) and, on the other, the integrals of interest that we shall evaluate will be found to be very large and strongly fluctuating functions of time (due to the near-singularity of the integrands), whereas the integral over the large semi-circle will be a constant, presumably of order unity. After writing out this integral in equation (10.10) and equation (10.11) , we shall ignore that integral by force of the arguments just presented.

10.3.3 Details of the model

Next we shall simplify and formalize the above model and shall describe what happens with the wp long before, long after, as well as at times close to collapse, bearing in mind the requirement of continuity in time. As already stated, we suppress dependencies on all coordinates (of the quantum system, the device and the surrounding) and let the label n in the wp sum stand for the various possible values in the measured property of the quantum system.

The stages that make up the wp history are the initial stage before the preparation of the wave-packet (I), the stage after preparation and before measurement (collapse) (II), the collapse-aftermath stage (III), ending up with the final stage (IV). For the values of the component coefficients in the initial stage c_n^I we postulate that the initial wp state is one of the pure states of the (whole) system which is in a thermal equilibrium (namely, one of the states of a canonical ensemble at a temperature T). After isolation of the system from its surrounding, the absolute values of the component coefficients will be close to the values $|c_n^T|$ representing thermal equilibrium,

$$|c_n| = |c_n^T| \equiv |c_n^I| \tag{10.5}$$

with some small random fluctuations around these values (which we shall ignore) and with the phases of c_n^I left free for the implementation of the future stages: preparation and measurement.

To apply this in a simple way, we assume in the numerical model that we shall work with, that there are in all 100 quantum states, which are supposed (for the sake of simplicity and democracy) to be all at the same energy-height. Thus, at stage I: $|c_n^I|^2 = \frac{1}{100}$ for all n. The wave packet in equation (10.4) is therefore properly normalized. (Other, more realistic assumptions for the numerical values in the model lead to the same conclusions as those with our simplistic choice of numbers.)

After preparation, at stage II, in our model 10 states are prepared and included in the wp, with 90 left outside the prepared state. Then the latter 90 have at this stage zero component weights, while the remaining 10, one $n = S$ to be selected and nine $n = NS$ not to be selected, have finite component weights. As, however, we insist on continuity, we let $log|c_n(t)|$ go continuously and *linearly* between the values at stage I and II, this change taking place during a time-duration Δt_p, starting from the initial time of preparation t_p (p for preparation). Formally, this means that

$$\log|c_n(t)| = \log|c_n^I| + \frac{t - t_p}{\Delta t_p}(\log|c_n^{II}| - \log|c_n^I|) \quad t_p \le t \le t_p + \Delta t_p \quad (10.6)$$

With the collapse starting at $t = t_c$ and coming to completion at $t_c + \Delta t_c$, we have similarly a continuous passage between stages II and III, shown formally by

$$\log|c_n(t)| = \log|c_n^{II}| + \frac{t - t_c}{\Delta t_c}(\log|c_n^{III}| - \log|c_n^{II}|) \quad t_c \le t \le t_c + \Delta t_c \quad (10.7)$$

The values after collapse (selection) are for the selected component $|c_S| = |c_S^{III}| = 1$ and for the nine unselected ones $|c_{NS}| = c_{NS}^{III} = 0$ (and the same value, of course, for the ninety components not prepared). Ultimately, starting at a thermalization time $t = t_T$, and becoming completed at $t = t_T + \Delta t_T$, all 100 component coefficients return (continuously, and in a manner similar to those between the earlier stages) in stage IV to their initial values given for stage I, namely

$$|c_n^{IV}| = |c_n^T| = |c_n^I| \quad all\ n \quad (10.8)$$

(Note that we are not postulating any special mechanism for thermalization. We do expect though that under fairly general condition a closed, but self-interacting system will converge to a superposition state which is similar to a thermalized state, in the sense that the components will be weighted with Boltzmann factors) The following scheme summarizes the time sequence of events with the stages marked underneath.

$$\begin{array}{ccccccc} -\infty & < & t_p < t_p + \Delta t_p < t_c < t_c + \Delta t_c < & t_T & < & t_T + \Delta t_T & < \infty \\ & (I) & (II) & & (III) & & (IV) \end{array} \quad (10.9)$$

Physically reasonable values for the times and time intervals are as follows (all values in s):

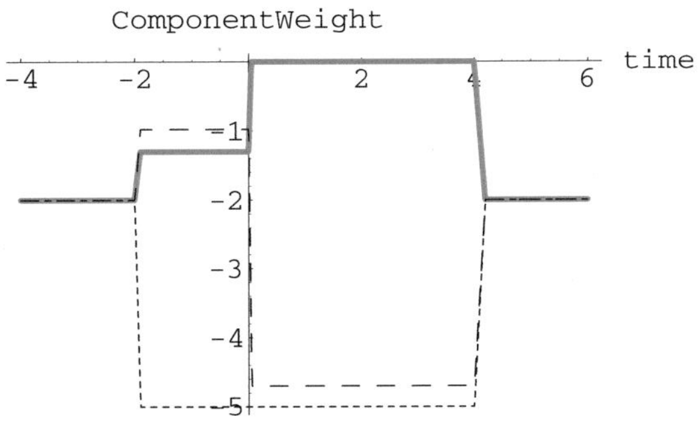

Figure 10.2: component weights as function of time in the model. (Schematic and scale-distorted for facility of presentation.) The decimal logarithms of the weights (moduli-squared) of the component coefficient s c_n are plotted against time. The selected component $n = S$ is shown by the thick line, a prepared but non-selected component $n = NS$ is drawn by long broken lines and a non-prepared component by short broken lines. Schematized times are $t \approx -4$ representing the system before preparation, when all components have the same weight; $t = -2$ instance of preparation, $t = t_c = 0$ time at which the measurement (collapse) begins; $t = t_T = 4$ at which thermalization begins. All processes are continuous but sharp, sudden. Component weights of order 10^{-4} or less signify zeros.

t_p (instance of wp preparation): -2
Δt_p (duration of preparation): 10^{-3}
t_c (time of initiation of collapse, or of measurement): 0
Δt_c(duration of collapse): 10^{-20*}
t_T (beginning of thermalization) : 4
Δt_T (thermalization time): 10^{-11}

[* Typical values of the time-duration of the collapse that have been proposed are indeed of the order of 10^{-20} seconds [26, 27]. (The times of these two references, though similar in value, are based on rather different mechanisms.)]

In the accompanying figure (Figure 10.2) we show the component moduli on a decimal logarithmic scale. We have made some adjustments in the parameters, so that the three types of moduli (of the selected (S), the non-selected (NS) and not-prepared components) can be separated visually and (again for purposes of presentation) we have also substantially increased the durations of the processes that take place between the subsequent stages.

10.4 The Component Phases

10.4.1 A Hilbert transform

Under the assumptions made for the analytic properties of the component-coefficients, the logarithm of the absolute modulus $\ln|c_n(t)|$ and the phase $\Phi_n(t) \equiv \arg\ c_n(t)$ of the n-component can be connected by a modification of two Hilbert transform relations [28]. Shown in [29], the appropriate relation for the phase is (temporarily suppressing the component index n)

$$\Phi(t) \ = \frac{1}{\pi} P \int_{-\infty}^{\infty} dt' \frac{\ln|c(t')|}{t'-t} - Et + \sum_j arg \frac{t-t_j}{t-t_j^*} + \Phi_\infty \tag{10.10}$$

Here P signifies the principal value of the singular integral and $\hbar = 1$ in the units we use.

The energy term, the Blaschke terms and the large arc integral

In the second term on the right we have the energy of the component. As a rule, this is numerically a very large quantity, since it represents an eigenenergy of a large sized body, which includes the device, etc. Fortunately, only the tiny part relating to the measured micro-system is of physical interest, as will be seen in section 10.5. We shall also see there that the effect of the first term is overwhelming. Therefore, we henceforth disregard the second, energy term. [The mathematical origin of this term in the transform formula is that a linear term in the phase of the amplitude does not satisfy the convergence requirements. So this part has to be taken away from the amplitude and then reinstated as in equation (10.10).]

The third term contains the Blaschke-phase sum [30] and involves the zeros t_j of $c(t)$ lying in the lower half of the t-plane (the one in which c(t) was assumed to be free of singularities), as well as the complex conjugates t_j^* of the zeros. The presence of this term can also be simply explained, along the lines of the energy term, as follows: For the Hilbert transform relations to be validly applicable to *log* $c(t)$, rather than to $c(t)$, one needs $c(t)$ to be free *also* of zeros in the lower half t-plane. We can assume that any zeros that may be are simple; higher order zeros require obvious adjustments in the expression. One can then easily eliminate the zeros of $c(t)$, by dividing this with factors of the form $t - t_j$ and multiplying it with $t - t_j^*$. The second operation is needed not to change the asymptotic behavior of $c(t)$, (assumed to be a constant for large t). The resulting function now satisfies the conditions needed for the Hilbert transformation. However, to obtain the phase of the original coefficient, we need to add the phase from the inverses of the factors by which $c(t)$ was changed. This is the third term in equation (10.10) .

The last term Φ_∞ stands for a line integral along the large semicircle $(C\ :\ t = |t_\infty|e^{i\theta}, |t_\infty| \to \infty, \theta = (\pi, 2\pi))$ in the lower half of the t-plane, defined and approximated by

$$\Phi_\infty = Re\frac{1}{\pi}\int_C dt' \frac{\arg c(t')}{t'-t} \approx -\frac{1}{\pi}\int_\pi^{2\pi} d\theta \arg c(|t_\infty|e^{i\theta}) \qquad (10.11)$$

As was argued in section 3.2, this is independent of t and of order unity, so that it can be neglected in the evaluation of the large component phase $\Phi(t)$.

Returning to the (first) term in the form of an integral in the above expression, one sees that this is zero for $|t| \to \infty$. By our postulate (which is really a convenient convention) that the phase of every component is zero initially, namely at $t = -\infty$, the Blaschke phase is also zero there and it can be assumed that this is true for each term in the Blaschke-sum. This means that we choose that branch of the multivalued arg's in the sum, which is zero at $t = -\infty$. With this choice, it can be shown that the Blaschke term is non-positive (negative, or zero when $c(t)$ has no zeros in the lower half t-plane). [To illustrate: let us a consider a single term with $t_j = -i$. so that the Blaschke-term is now $arg\frac{t+i}{t-i} = -2arctan\frac{1}{(-t)}$. This can be readily seen to take the following values: For $t = -\infty$ the argument is 0, as postulated; for $t = -1$, it is $-\frac{\pi}{2}$; for $t = 0$, it is $-\pi$. Lastly, for $t \to \infty$ it tends to -2π. Thus this term, and its time derivative, is non-positive throughout and so is the sum of similar terms.]

With this in mind, we can calculate the *upper*-limit of the phase of each $c(t)$ from the integral (the first term in equation (10.10)) and this requires as the input only the values of $|c(t)|$ for real times. For brevity and for reasons that will become clear, in what follows we shall refer to the contribution to the phase from the integral as *the* phase.

10.4.2 Expression for the phase

Inserting into the integral the modulus values for the model in section 10.3, we obtain a closed expression for the phase. An important mathematical fact in the analysis that has led to the expression to follow is that if $\ln|c(t')|$

is constant in the integrand of equation (10.10) , the integral is zero. This means that in the phase we shall find that all coefficient moduli are expressed as relative to (=divided by) their long-term, or equilibrated values, namely those at stages I and IV. The introduction of this *non-zero*, equilibrated long-term value for the component coefficients is one of the advances made in this work and removes difficulties encountered in earlier uses of Hilbert transforms for the "phase-problem" [31]. (Had we have $c_{NS} -> 0$, this would make the $\log c_{NS}$ integration singular. As a more recent application of Hilbert transforms to geometric phases we note [32].)

After some simplification, we obtain for the phase $\Phi_n(t)$ of the n-component coefficient

$$\begin{aligned}\Phi_n(t) &= \frac{1}{\pi}\ln\frac{c_n^p}{c_n^T}[\frac{t-t_p}{\Delta t_p}\ln\frac{|t-t_p-\Delta t_p|}{|t-t_p|}+\ln\frac{|t-t_c-\Delta t_c|}{|t-t_p-\Delta t_p|} \\ &- \frac{t-t_c}{\Delta t_c}\ln\frac{|t-t_c-\Delta t_c|}{|t-t_p|}] \\ &+ \frac{1}{\pi}\ln\frac{c_n^c}{c_n^T}[\frac{t-t_c}{\Delta t_c}\ln\frac{|t-t_c-\Delta t_c|}{|t-t_c|}+\ln\frac{|t-t_T-\Delta t_T|}{|t-t_c-\Delta t_c|} \\ &- \frac{t-t_T}{\Delta t_T}\ln\frac{|t-t_T-\Delta t_T|}{|t-t_T|}] \qquad (10.12)\end{aligned}$$

This expression is the basis of our analysis of the phase acquired by the three types of components: selected ($n = S$), non-selected (NS) by the measurement and not-prepared (NP) by the wp formation process. The crucial difference between these is that the logarithms of the amplitude-moduli of the selected component $\ln|c_S|$ are finite at all stages, whereas $\ln|c_{NS}|$ for a non-selected component tends to $-\infty$ immediately after the collapse (denoted by the superscript c) and before the thermalization stage (since $|c_{NS}^c| \to 0$). The non-prepared components have vanishing amplitudes at all except the initial and final thermalized stages. In our analysis of the expression, we shall let the vanishing amplitude approach zero as a limiting process, rather than actually let it take a zero value, which would give a singularity.

Another quantity whose limiting value will be of interest is the collapse "aftermath time". This is denoted, as before, by t_T, and represents the time it takes for the component amplitudes to start to become thermalized or equilibrated.

Let us emphasize, that we expect that the true, physical amplitude-moduli will be different and much more complicated than the modelled one, since the microscopic system is under interaction with a macroscopic measuring device comprising sensors, pointers, etc. Therefore the functional form of the true amplitudes is bound to be different. However, the basic nature of the processes (preparation, collapse, thermalization) makes it legitimate to assume that the numerical values of the component-moduli on the real time-axis will be similar to the postulated ones. If so, then the values of the phase calculated from the integral relation with the model component-moduli will yield approximately the phases of the true amplitudes. This remains true (since, as noted, the integral relation requires the modulus values only on the real axis), no matter how wrong the postulated values are *away* from the real axis or that the coefficients *of the model* do not possess the analyticity properties assumed.

Evaluation of the phases

We shall first concentrate on the phase that the formula predicts for a non-selected component.

NS : Using a simple limiting procedure, one can verify that in the two extremes, as $|t| \to \infty$, the phase vanishes. We next consider what happens at the beginning of collapse, at the moment that $t = t_c$. Because $|c^c_{NS}|$ is vanishingly small, it is the terms in the third and fourth lines in equation (10.12) that dominate (logarithmically and *negatively*) the phase, and do so through the second logarithmic term inside the second square bracket. The second logarithm inside the square brackets can be seen to be positive, so that the phase acquired at this moment is negative. It is also numerically large, due to the large value of $|log|c^c_{NS}||$.

We now examine the dependence of the phase on the thermalization time t_T (which we call "the collapse aftermath"). Some time after collapse and well before thermalization is reached,

$$t_c < t << t_T \tag{10.13}$$

as t_T takes on large values, the phase diverges negatively and linearly with $\ln t_T$. This follows from the second term in equation (10.12) , where again the second, logarithmic term dominates. The behavior of the NS phase, as a joint function of the non selected amplitude and of the collapse aftermath time, is shown in figure 10.3(a).

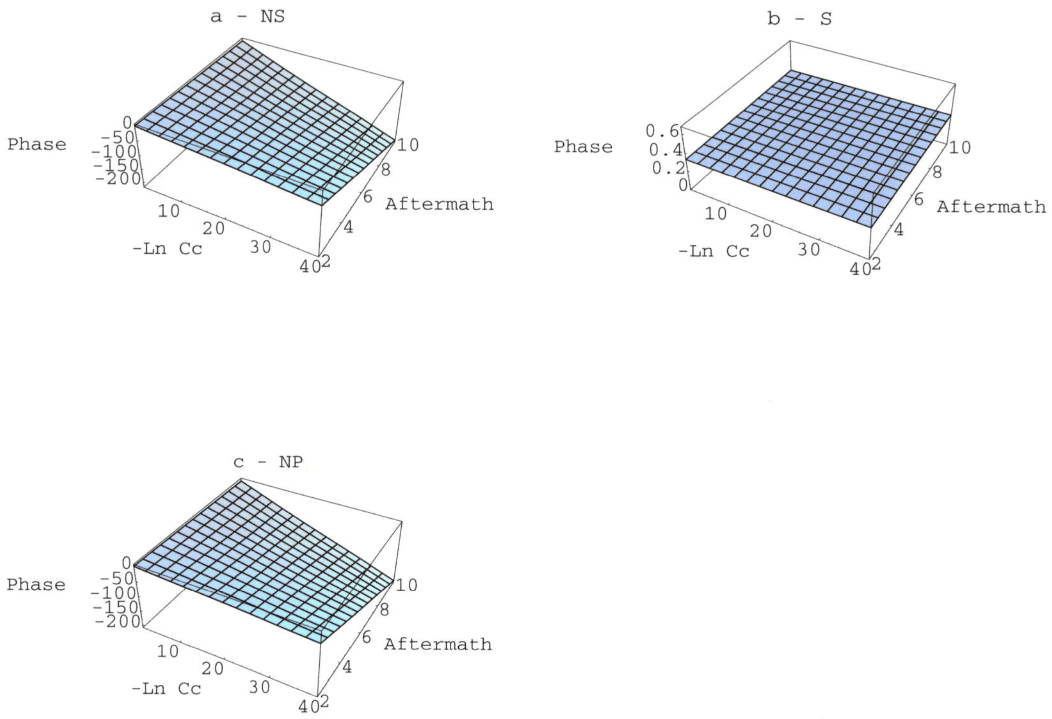

Figure 10.3: Component phase: asymptotic dependence on collapse aftermath duration (after collapse time t_c and before reaching thermalization at time t_T, as in the text) and on the value of component modulus c^c after measurement (a) for a non-selected (NS) component. (b) for the selected (S) component, and (c) for a non-prepared component. The phase is calculated at 1 s after the measurement, this time being just quarter-way between t_c and t_T.

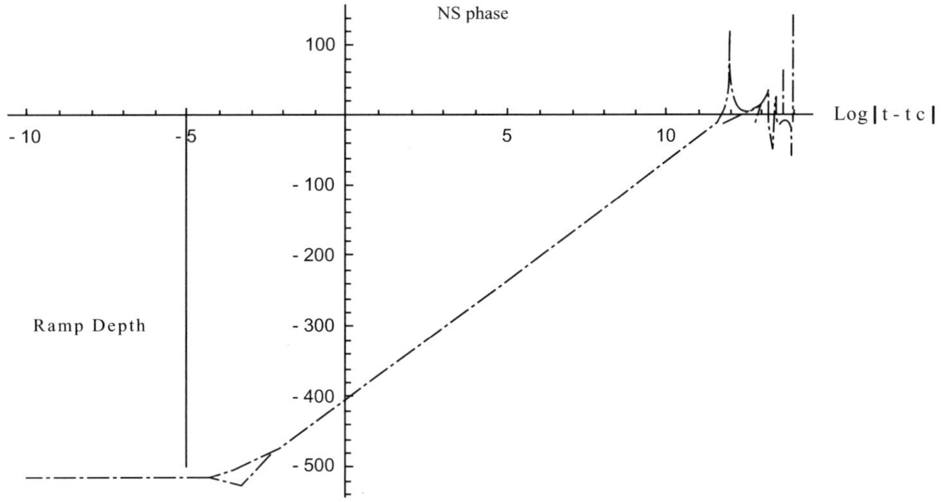

Figure 10.4: Dependence of a NS component on the time difference from measurement t_c, expressed as the decimal logarithm of $|t - t_c|$. The phase before the beginning of the measuring process is shown by broken lines and after it by the dash-dot curve. Note that the right hand extremity of the figure is for times long before and long after the collapse and the left hand extremity shows component moduli for times very close to the time of measurement. On the left the phase scales with $|\ln c^c_{NS}|$ and tends (logarithmically) to -∞ for vanishing $|c^c_{NS}|$

In the figure the value of the phase (when expressed in radians) lies in the hundreds. Though the plotted phase is not infinite, the logarithmic divergence with the aftermath duration and with the value of component modulus after measurement is apparent.

It is now also clear why the so far disregarded Blaschke phase (arising from zeros of the component function) is not important. This is because this non-positive term can only further lower the negative phase, but cannot change its tendency. The other two pictures in Figure 3 illustrate the behavior of the selected S component [Figure 10.3(b)] and of a component not included in the wp [Figure 10.3 (c)]. The flatness and small magnitude of the phase for the selected component are worthy of notice.

The accompanying figure (Figure 10.4) shows a NS-phase as function of the time difference from the initiation of the measurement process, both before and after the event. The broken curve shows the phase-history before and the broken-dotted curve after the measurement. Thus time runs on the two curves in opposite directions. Insomuch that time is shown as the (10-

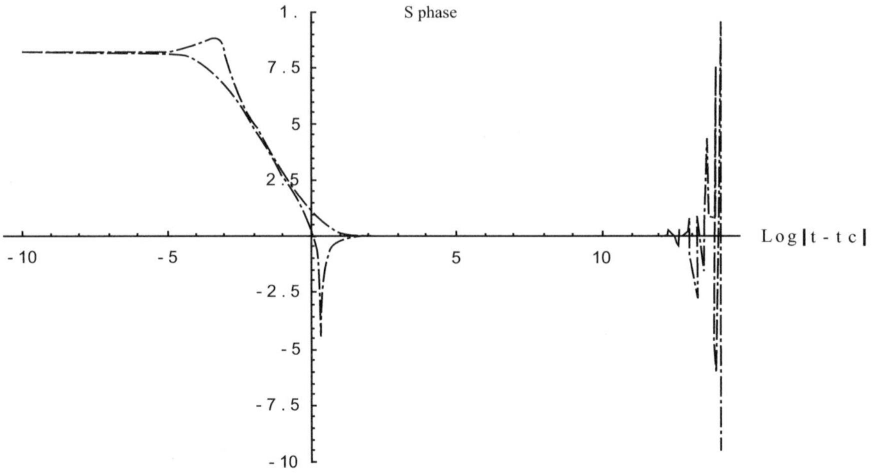

Figure 10.5: Dependence of the selected S component on the time difference from measurement. The remarks in the caption of figure 10.4 apply here, too. Additionally, the next two figures illustrate the first and second derivatives of the NS phase. (Figures 10.6 and 10.7)

based) logarithm of $|t - t_c|$, the breadth of the left hand part is deceptive; it really represents a relatively short epoch. The ramp-depth is the measure of the acquired phase and it is this quantity which is proportional to the log of the component modulus immediately after the collapse. The rough identity of the phase before and after the collapse time t_c is apparent. It is perturbed only by events representing changes *in the slope* of the component moduli, as given in the description of the model in section 10.3. They are of minor importance and are model dependent, in the sense that if we had let the time- *derivatives* of the component amplitude be continuous at t_c, then the peaks would disappear. Notable is, however, that the phase senses the measurement process already *before* the collapse. This is a consequence of the analytic behavior of the component. The phenomenon was discussed before in the context of causality [33] and, though it is an interesting issue, we will touch on it only briefly in the concluding section.

For comparison we present the phase of the selected (S)-component (figure 10.5). At times very close to collapse (the left hand side of the figure) this is about two orders of magnitude less than the absolute value of the phase of a NS component (shown in the previous figure).

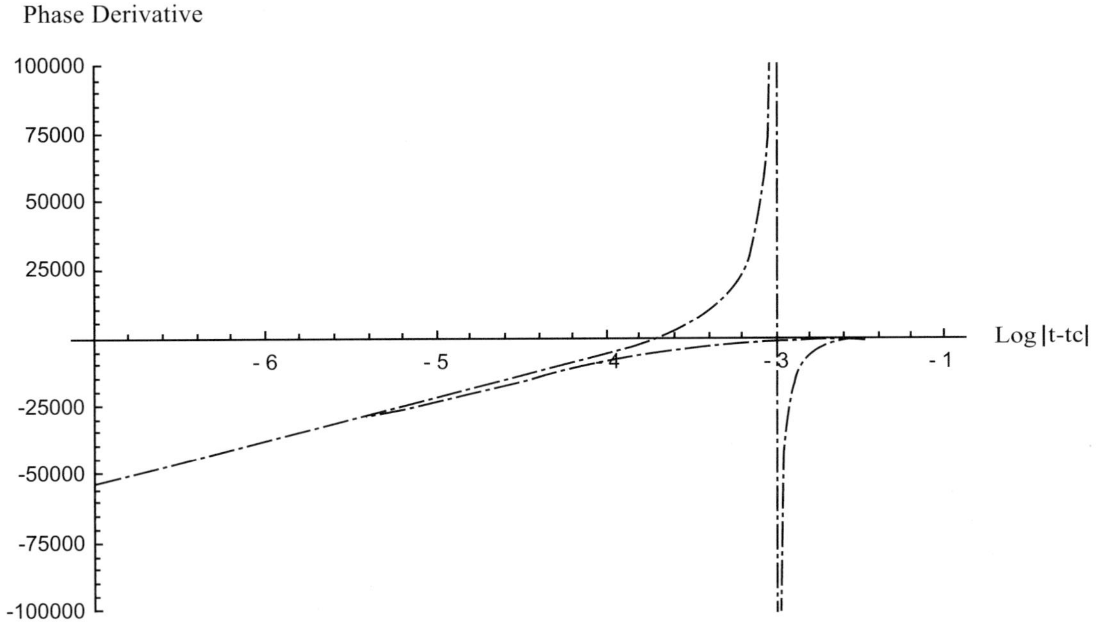

Figure 10.6: Time derivative of the phase of a NS-component. (See caption of figure 10.4.)

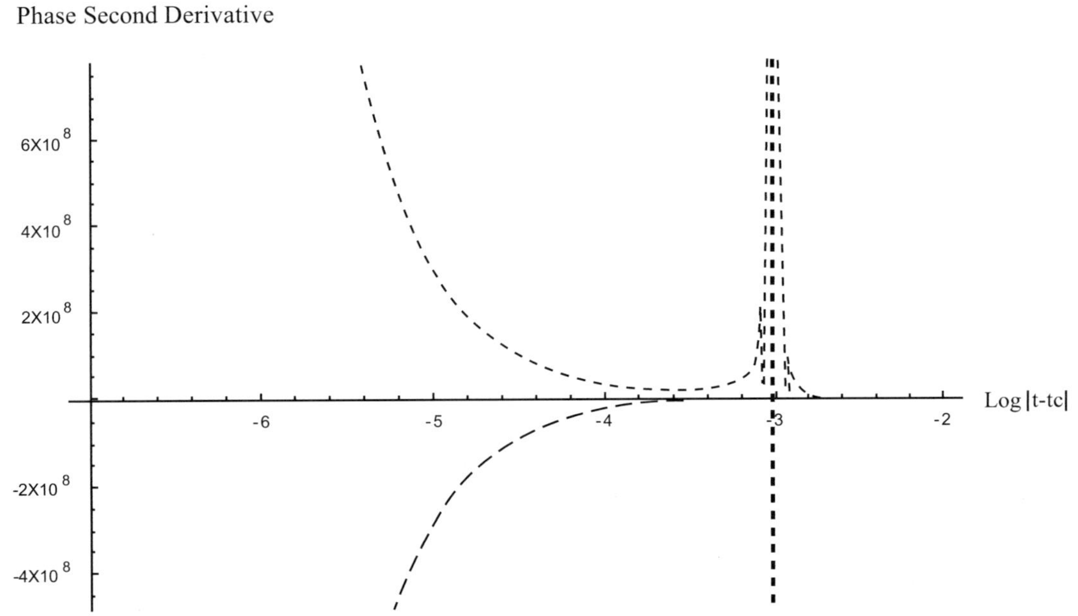

Figure 10.7: Second time derivative of the phase of a NS-component. (Here, too, the caption of figure 10.4 is helpful.)

10.5 Removal of a fast oscillating component.

We have found that, subject to the mathematical assumptions made in section 10.2, the phase $\Phi_{NS}(t)$ in each disappearing component amplitude $c_{NS}(t)$ becomes in a wp - collapse exceedingly large (in its absolute value), proportional to the logarithm of the "vanishing" NS amplitude. The very large phase is tantamount to a very rapid oscillation of the phase factor, similar to the so called *Zitterbewegung* in the velocity of a Dirac-electron. The effective vanishing of a fast oscillating term was suggested on p. 263 in [34]. We now wish to formally demonstrate the removal of rapidly varying component from the wp, if only in a non-rigorous way.

Let us postulate that a component can be said to be "removed", if it cannot be detected by any emission or absorption of electromagnetic radiation or any other mode of interaction. In the case of light, the transition amplitude A that has to be non-zero for emission or absorption of a photon quantum $\hbar\Omega$ is

$$A(\Omega) = M \int_{-\infty}^{\infty} dt |c_{NS}(t)| e^{i(\Phi_{NS}(t) - \Omega t)} \tag{10.14}$$

Here M is a matrix amplitude that mediates the transition to some other state of the measured quantum system and includes also photon amplitudes as well as wave function amplitudes of the environment which do not change during the transition. Since we are interested in the effect of only the phase factor on the transition probability, we shall take out the modulus $|c_{NS}(t)|$ from under the integration and incorporate it into the prefactor M.

An estimation of the remaining integral can be obtained by the stationary phase method to yield approximately

$$A(\Omega) \approx NMe^{i(\Phi_{NS}(t_{SP}) - \Omega t_{SP})} \left| \frac{d^2\Phi_{NS}(t_{SP})}{dt_{SP}^2} \right|^{-\frac{1}{2}} \tag{10.15}$$

Here t_{SP} is the stationary phase value of the integrand. N is numerical, equal to $(1 \pm i)\sqrt{\pi}$ (where the sign follows the sign of the second time derivative of Φ_{NS}) and the exponential has the absolute value of unity. In the above expression it is the second derivative between the modulus bars, which ensures the asymptotic vanishing of the transition probability: in the neighborhood of t_c (the collapse time), where the phase is (numerically) large, the second derivative is of the order of

$$\frac{d^2\Phi_{NS}(t_{SP})}{dt_{SP}^2} \approx |lnc_{NS}^c| (\Delta t_c)^{-2} \tag{10.16}$$

We find therefore that the transition probability $|A(\Omega)|^2$ vanishes logarithmically with the vanishing of the non-selected component amplitude. Note also that this, asymptotic value of the transition amplitude is proportional to the square of the duration of the collapse process Δt_c and is independent of the duration of the aftermath.

By consequence of this result, the non-selected states cannot be detected electromagnetically, entirely due to their phase divergence. Similar conclusion follows for absorption or emission of other bosons (e.g. material vibrations). Other signatures of a component may, in principle, also be possible but are not evident to us. A similar conclusion was reached by a different route in [35].

How is this result affected by the three terms on the right hand side of equation (10.10) , that have been neglected? One can disregard the last term, which is a constant. The energy term, linear in time, is also immaterial since its second derivative is zero. The same is true for those terms in the Blaschke sum, for which the zeros t_j lie far from the origin, or the time of measurement, in the sense that $|t_j| >> \Delta t_c$. Regarding those zeros which are close to the origin, it is not possible to estimate their contribution to the second derivative without knowing the precise analytic form of the amplitude $c_{NS}(t)$. Still, supposing as a worst case scenario, that their number is of the order of $|lnc^c_{NS}|$ or higher, so that their contribution is comparable to the quantity in equation (10.16) , it is extremely unlikely that they will precisely cancel. (We recall that it is the absolute value of the second derivative that appears in equation (10.15)).

10.5.1 Density matrix treatments

We have thus demonstrated the effective removal of the NS terms from a wp, by a phase factor cancellation mechanism. The corresponding removal of fast oscillating off-diagonal elements in the density matrix has been the core of many phase-decoherence theories [4]. In this brief part we cite results from just two explicit calculations of the off-diagonal part in the density matrix of the system, upon which a measurement is performed. These also show the simultaneous presence of damping and of phase divergence in a collapse, but higher order time-derivatives of the phase do not show the divergence that we have found for the NS components in wp's.

The Jaynes-Cummings model

This model involves two interacting quantum systems: an atom with two levels ($|+>$ and $|->$) and a single mode of the quantized electromagnetic field, the interaction between which (characterized by a strength g) is expressed in the rotating wave approximation. In an exact treatment of this model, collapse was treated by showing the vanishing of the off-diagonal element of the "atom density matrix" $\rho_{at}(t)$ [36]. The combined system is initially prepared in a state

$$\Psi = (\gamma|+> + \delta|->)|v\} \tag{10.17}$$

where γ, δ are constants and $|v\}$ is the initial coherent state of the field. The off-diagonal element of the partial, "atomic" density matrix is calculated in [36] and has the form (shown in Eqs. 18 of that work)

$$< +|\rho_{at}(t)|- >= \gamma\delta^* e^{-igt\sqrt{\bar{n}}}\{ve^{\frac{igt}{2\sqrt{\bar{n}}}}|ve^{-\frac{igt}{2\sqrt{\bar{n}}}}\} \tag{10.18}$$

where $\bar{n}$ denotes the mean number of field oscillators $\approx v^2$. The complex matrix element can be evaluated using the formula for coherent states

$$\{u|w\} = \exp[-\frac{|u|^2 + |w|^2}{2} + u^* w] \tag{10.19}$$

(Eq. 19 in [36].) Substituting this into equation (10.18) , one obtains, apart from terms that are time independent, the modulus and the phase of the off-diagonal matrix element, as

$$| < +|\rho_{at}(t)|- > | \quad = \quad \exp[-v^2(1 - \cos\frac{gt}{\bar{n}})] \approx e^{-\frac{g^2 t^2}{2}} \tag{10.20}$$

$$arg < +|\rho_{at}(t)|- > \quad = \quad -gt\bar{n} - v^2 \sin\frac{gt}{\bar{n}} \approx -2gt\bar{n} \tag{10.21}$$

The approximations made in the last two equations are those in [36], namely truncating the analysis at times shorter than $\frac{v}{g}$, but longer than $\frac{1}{g}$. It is evident that in the classical limit $v^2 \approx \bar{n} >> 1$ not only is the modulus strongly damped, but also the phase increases (numerically) very fast. One thus finds a simultaneous damping and phase divergence, whose source (for both) is the macroscopic nature of the field. However, the second time-derivative of the phase is finite. (We also note in passing, that the modulus and phase of the matrix element satisfy equation (10.10) , with both the Blaschke-term and the last, large semicircle-integrand being zero. This indicates, but does not prove, that the analyticity conditions postulated in section 10.3 are satisfied by the off-diagonal matrix element, as a function of the complex time.)

An energy driven collapse model

A later work ([17], which gives references to earlier studies) investigates mechanisms of collapse in which the system "homes onto" one of its energy eigenvalues, rather than to a position value (of, say, a pointer). Explicit expressions for the off-diagonal matrix elements of the density matrix are derived, also within a many-universe treatment. The homing process involves a diffusion in energy space (more accurately in inverse energy space) whose speed is given by a diffusion constant λ. With some simplification the environment averaged matrix element in a precessing two-state model can be written (cf. eq. (6.8) in [17]) as

$$< +|\rho(t)|- >= Ce^{-\epsilon^2 \lambda t/2 - i\epsilon t} \tag{10.22}$$

where C is constant. The limit of effective damping-out of the off-diagonal term $\epsilon^2 \lambda t >> 1$ requires for slow diffusion a very high energy value, or a fast varying phase of the matrix element, in accord with the present chapter's claim. The second time derivative of the phase is, however, zero, in contrast to what we have found for the wave packet models.

10.6 Global estimates

The following well known integral relation holds between Hilbert transforms [28] and can provide a further estimate for the phase, without resorting to computation (but still needing the assumptions about analytic behavior).

$$\int_{-\infty}^{\infty} dt (ln|c(t)|)^2 = \int_{-\infty}^{\infty} dt (arg[c(t)])^2 \tag{10.23}$$

The left hand side can easily be estimated, utilizing the constructions in the previous section or the drawings in Fig. 10.2. We aim at obtaining estimates for the arg (or phase angle) appearing on the right.

For the component S selected by the measurement, the integration in equation (10.23) is of a finite range (between t_p and $t_T + \Delta t_T$] and the integrand is finite throughout (since $|c_s|/|c_s^T|$ never vanishes). Thus, on the right hand side of equation (10.23) the argument (the phase) is always finite.

In contrast to this, for a non-selected NS component, in the left hand member of the above relation the modulus is zero in the aftermath of the measurement until thermalization. The right hand member is therefore a divergent quantity times "the length of the aftermath". The right hand part is therefore also divergent, but it is not clear whether this is due to a divergence in the integrand (the squared phase) or to the infinite range of the integral. To decide between the alternatives, we take the NS-moduli to be very small, but non-zero after the measurement. Then we see from the original integral relation in equation (10.10) , which we recall involves only a finite integration range, that the phase decreases with time (for real t) at least as fast as $\frac{1}{|t|}$. This means that in the right hand member of equation (10.23) the integration converges at the infinite limits. Therefore, the divergence that arises in this integral as the non-selected modulus tends to zero (and its logarithm becomes $-\infty$) is due to the phase itself diverging along the real time axis.

We conclude from this exercise that (subject to the analytic conditions postulated for the component amplitudes), the vanishing of the NS - amplitude implies a logarithmically infinite phase. The same result holds also for components that are not included in the prepared wave packet.

10.7 Bohm's Description of Decoherence

In one of the earlier scenarios of wave-function collapse in terms of phase-decoherence (equated by him to "destruction of interference") D. Bohm gave a semi-quantitative account of this decoherence, exemplifying it with the cases of an electron just having emerged from two slits or having passed a Stern-Gerlach apparatus reaching positions A and B on a screen [8].(The pioneering content of his account may be gauged from the fact that he did not fail to mention gravity as a possible cause of the collapse.) He wrote [with our additions]: "Thus the phase of the wave function [component] is made completely indefinite and there is no deterministic relation between the phase before the interaction [with the measuring apparatus] and the phase after the interaction. Furthermore, ...there is no definite relation between α_A [the phase arising from the wave emanating from slit A] and α_B so that the phase difference $\alpha_A - \alpha_B$ is also unpredictable and uncontrollable [as well as extremely large, $\approx 10^{11}$ in an example given in by D. Bohm in section 22 of [8]].

Because the phases ... fluctuate in a random and uncontrollable way from experiment to experiment [not everyone would endorse the necessity of the last four words, which can probably be traced to Bohm's dialectical materialistic philosophy, that "probability" has a meaning only insofar as it refers to a set of similar experiments carried out under equivalent initial conditions] terms like $e^{i(\alpha_A - \alpha_B)}$ will average out to zero."

On the one hand Bohm presents us with a credible decoherence scenario arising from large component-phases (e.g., α_A) along the lines envisaged in the present work, but he also predicts the same fate for the *relative* phases $\alpha_A - \alpha_B$. But it is just these phase difference that appear in the off-diagonal elements of the density matrix. We thus arrive to the same applicability of the present amplitude-phase-interdependence theory to both alternative descriptions, noted in the introduction, namely those anchored either in NS-phases or in the off-diagonal elements of the density matrix.

We further note that in Bohm's description there exists a causal link from the observing apparatus and the interference destruction, while in the present article decoherence is concomitant with a state selection, S-YES, NS-NO process. But, of course, such state selection occurs only through the agency of the measuring apparatus. Thus, our formalism, which does not explicitly include the measuring apparatus, contains its effect on the observed system.

In descriptions differing from the present one, in which the environment is described by quantum variables and the full state is the entangled state of the system and the environment, the violent fluctuations in the system wave function are the stamps of the environmental wave function. The full wave function (representing a closed system) will not fluctuate, but the component fluctuations will force the environment into one of its pointer states, as described in [37] and by several authors subsequently.

A recent blog [38] refers to the very recent directions in collapse theories and demands that the kinematical aspect (i.e., the identification of the preferred pointer basis) should not diminish the importance of the dynamic aspect, which is the emergence of the decoherence (due to an external agent) or collapse.

10.8 Conclusion

We have found that, subject to the mathematical assumptions made in section 10.2, the *phase* of each non-selected (NS) component amplitude becomes in a wp - collapse exceedingly large (in its absolute value) and this happens very fast. Two circumstances contribute to the large magnitude of the phase: the very fast collapse $\Delta t_c \approx 10^{-20}$ s and the long period (t_T) during which the

NS components are removed from the wp. (t_T can take up a wide range of values, around a second, after which the components are deemed to return to the equilibrium values that they had before the wp preparation procedure.)

In addition, the fast acquisition of a large phase is tantamount to a rapid variation in the phase factor and we have seen in section 5 that this variation can lead on its own to a removal of the component, without needing the vanishing of the component modulus. In a real collapse both events take place, due to the mathematical interconnectedness of phase and modulus. In the introduction we have noted several well-known instances which show that the connection is particularly dramatic, when the log of the component modulus becomes singular. This is of course the case in a collapse process.

Thus, the two mechanisms of collapse, amplitude decay as in [13] and decoherence as in [12] are intertwined and one mandates the other. Like many effects previously found ([6],[14]), this manifestation of an interrelation between the phase and modulus (or more precisely its logarithm) is inherent in the time dependent Schrödinger equation.

Reduced density matrix formalisms which are descriptive of decoherence mechanisms also show that the macroscopic nature of the environment enters both the moduli and the phase of off-diagonal elements (which is essentially the thesis of this work), but the analytic form of the time-dependence is not (to our mind) firmly established and needs further work.

In section 10.2.1 it was found that the enhancement of the negative phase in a NS component starts at some time (of the order of the small quantity $\Delta t_c \approx 10^{-20}\ s$) *before* the "time of the measurement" t_c. Though an apparent violation of causality occurs, in truth this is a logically necessary outcome of our having assumed analytic behavior for the wp (and this has been partly justified in this work in relation with a closed system). This was also the conclusion of a discussion of a similar effect, which appeared in the context of causality- superluminality [33]. Further treatment of this interesting topic would involve the questions whether this paradox can be (or need be) resolved in a practical situation and how it relates to quantum non-locality? These questions may be left to a later work.

Acknowledgement

We thank Larry Horwitz for a constructive reading of an earlier version of our manuscript and a discussion.

Bibliography

[1] J.W.S. Rayleigh (Lord Kelvin), The Theory of Sound. Dover: New York, 1945. Volume II, Section 282.

[2] L.Mandel and E.Wolf, Optical Coherence and Quantum Optics. University Press: Cambridge, UK, 1995. Section 3.1.

[3] E.P. Wigner, Philosophical Reflections and Syntheses. Springer-Verlag: Berlin, 1997.

[4] D. Giulini, E. Joos, C. Kiefer, J. Kupsch, I.O. Stamatescu and H.D. Zeh, Decoherence and the Appearance of a Classical World in Quantum Theory. Springer: New York, 1996.

[5] H.D. Zeh in [4] p. 5.

[6] I.O. Stamatescu in [4] p. 249.

[7] G.W. Ford, J.T. Lewis, and R.F. O'Connell, "Quantum Measurement and Decoherence," *Phys. Rev. A*, vol. 64, 032101, 2001.

[8] D. Bohm, Quantum Theory. Prentice Hall: New Jersey, 1951. Particularly sections 6.4, 22.8-9

[9] J. von Neumann, The Mathematical Foundations of Quantum Mechanics. Dover Books: New York, 1959.

[10] G. Ghirardi, A. Rimini, and T. Weber, "Unified dynamics for microscopic and macroscopic systems," *Phys. Rev. D* , vol. 34, 470, 1986.

[11] A. Stern, Y. Aharonov, and Y. Imry, "Phase uncertainty and loss of interference: A general picture," *Phys. Rev. A*, vol. 41, 3436, 1990.

[12] R. Omnes, "Decoherence, irreversibility, and selection by decoherence of exclusive quantum states with definite probabilities," *Phys. Rev. A*, vol. 65, 052119, 2002.

[13] S.L. Adler, "Environmental influence on the measurement process in stochastic reduction models," *J. Phys. A: Math. Gen.*, vol. 35, 841, 2002.

[14] R. Englman, A. Yahalom, and M. Baer, "Time-dependent and time-independent approaches to study effects of degenerate electronic states," *J. Chem. Phys.*, vol. 109, 6550, 1998.

[15] K.G. Kay, "Integral expressions for the semiclassical time-dependent propagator," *J. Chem. Phys.*, vol. 100, 4377, 1994.

[16] W.H. Zurek, "Environment-induced superselection rules," *Phys. Rev. D*, vol. 26, 1862, 1982.

[17] P. Pearle, "Problems and aspects of energy-driven wave-function collapse models," *Phys. Rev. A*, vol. 69, 042106, 2004.

[18] A. Albrecht, "Following a "collapsing" wave function," *Phys. Rev. D*, vol. 48, 3768, 1993.

[19] M. Kac, "On the distribution of values of trigonometric sums with linearly independent frequencies," *Am. J. Math.*, vol. 65, 609, 1943.

[20] J. Friedel, "The distribution of electrons round impurities in monovalent metals," *Phil. Mag.*, vol. 43, 153, 1952.

[21] R. Englman, and A. Yahalom, "Complex States of Simple Molecular Systems," In: The Role of Degenerate States in Chemistry, editors M. Baer, and G.D. Billing, Adv. Chem. Phys., vol. 124, 197 section VI B, John Wiley and Sons: New York, 2002.

[22] L. Khalfin, "Contribution to the decay theory of a quasi-stationary state," *Soviet Phys. JETP*, vol. 6, 1053, 1958.

[23] S.I. Tomonaga, Quantum Mechanics, Vol II. North Holland: Amsterdam, 1966. Sections 41 and 61.

[24] E.J. Heller, "Frozen Gaussians: A very simple semiclassical approximation," *J. Chem. Phys.*, vol. 75, 2923, 1981.

[25] R. Englman and A. Yahalom, "Reciprocity between moduli and phases in time-dependent wave functions," *Phy. Rev. A*, vol. 60, 1802, 1999.

[26] P. Pearle, and E. Squires, "Bound state excitation, nucleon decay experiments and models of wave function collapse," *Phys. Rev. Lett.*, vol. 73, 1, 1994.

[27] R.P. Feynman, F.B. Moringo and W.G. Wagner, Lectures on Gravitation. Cal. Tech. Press: Pasadena, 1973.

[28] E.C. Titchmarsh, Introduction to the Theory of Fourier Integrals. Clarendon Press: Oxford, 1948. Chapter V.

[29] L. Mandel and E. Wolf, Coherence and Quantum Optics. University Press: Cambridge, 1995. Section 7.3.2.

[30] N.G. van Kampen, "S-Matrix and Causality Condition. I. Maxwell Field," *Phys. Rev.*, vol. 89, 1072, 1953.

[31] R.E. Burge, M.A. Fiddy, A.H. Greenway, and G. Ross, "The Phase Problem," *Proc. Roy. Soc. London A* , vol. 350, 191, 1976.

[32] A. Botero, "Geometric phase and modulus relations for probability amplitudes as functions on complex parameter spaces," *J. Math. Phys.*, vol. 44, 5279, 2003.

[33] D. Rohrlich and Y. Aharonov, "Cherenkov radiation of superluminal particles," *Phys. Rev. A*, vol. 66, 042102, 2002. Section V.

[34] P.A.M. Dirac, The Principles of Quantum Mechanics, Fourth Edition. Clarendon Press: Oxford, 1958. Section 69.

[35] M. Frasca, "Dynamical disappearance of superposition states in the thermodynamic limit," *J. Phys. B: At. Mol. and Opt. Phys.*, vol. 36, 2757, 2003.

[36] J. Gea-Banacloche, "Atom- and field-state evolution in the Jaynes-Cummings model for large initial fields," *Phys. Rev. A*, vol. 44, 5913, 1991.

[37] D. Bohm, "A Suggested Interpretation of the Quantum Theory in Terms of "Hidden" Variables. II," *Phys. Rev.*, vol. 85, 180, 1952. Section 2.3.

[38] G. Bacciagaluppi, "The Role of Decoherence in Quantum Mechanics," In: Stanford Encyclopedia of Philosophy, 2007.

Index

Made in the USA
Las Vegas, NV
03 November 2026

57424411R00178